DESIGNING SAFE ROAD SYSTEMS

Human Factors in Road and Rail Transport

Series Editors

Dr Lisa Dorn
Director of the Driving Research Group, Department of Human Factors, Cranfield University

Dr Gerald Matthews
Professor of Psychology at the University of Cincinnati

Dr Ian Glendon
Associate Professor of Psychology at Griffith University, Queensland, and is president of the Division of Traffic and Transportation Psychology of the International Association of Applied Psychology

Today's society confronts major land transport problems. Human and financial costs of road vehicle crashes and rail incidents are increasing, with road vehicle crashes predicted to become the third largest cause of death and injury globally by 2020. Several social trends pose threats to safety, including increasing vehicle ownership and traffic congestion, advancing technological complexity at the human-vehicle interface, population ageing in the developed world, and ever greater numbers of younger vehicle drivers in the developing world.

Ashgate's Human Factors in Road and Rail Transport series makes a timely contribution to these issues by focusing on human and organisational aspects of road and rail safety. The series responds to increasing demands for safe, efficient, economical and environmentally-friendly land-based transport. It does this by reporting on state-of-the-art science that may be applied to reduce vehicle collisions and improve vehicle usability as well as enhancing driver wellbeing and satisfaction. It achieves this by disseminating new theoretical and empirical research generated by specialists in the behavioural and allied disciplines, including traffic and transportation psychology, human factors and ergonomics.

The series addresses such topics as driver behaviour and training, in-vehicle technology, driver health and driver assessment. Specially commissioned works from internationally recognised experts provide authoritative accounts of leading approaches to real-world problems in this important field.

Designing Safe Road Systems

A Human Factors Perspective

JAN THEEUWES
Vrije Universiteit Amsterdam
RICHARD VAN DER HORST
Netherlands Organisation of Applied Scientific Research (TNO)
MARIA KUIKEN
DHV, The Netherlands

CRC Press
Taylor & Francis Group
Boca Raton London New York

CRC Press is an imprint of the
Taylor & Francis Group, an **informa** business

CRC Press
Taylor & Francis Group
6000 Broken Sound Parkway NW, Suite 300
Boca Raton, FL 33487-2742

First issued in paperback 2017

Version Date: 20160226

ISBN 13: 978-1-4094-4388-9 (hbk)
ISBN 13: 978-1-138-07376-0 (pbk)

Visit the Taylor & Francis Web site at
http://www.taylorandfrancis.com

and the CRC Press Web site at
http://www.crcpress.com

Contents

List of Figures

List of Tables

Foreword

Ton van den Brink and Michel Lambers

Rijkswaterstaat, the Dutch Road Authority, has worked with experts on human factors for a long time. Particularly concerning road safety, research was mainly focused on human factors. Dare say, this has made a tremendous contribution to the high level of traffic safety in the Netherlands.

Nowadays, Dutch experts on road design and traffic management recognize that knowledge of human factors is insufficiently used to reach their goal: an increase in road throughput. The main motivation for this goal is the social importance of a reliable throughput, even though there is steadily increasing traffic volume on the roads.

As a result of this *Rijkswaterstaat* is focused on network management and has put a lot of effort and money into ways of improving the throughput of infrastructural networks, with a focus on the desired behaviour of the road user.

Too often it turns out that individual or even large groups of road users can behave in a way that is undesirable for throughput or safety. Sometimes this is due to factors that are a social problem, such as speeding, alcohol or drug use or any other kind of deliberate disobedience of traffic rules. But there are also many other factors involved – such as comprehensibility (e.g. of traffic signs), task complexity (e.g. performing the driving task) or willingness (e.g. to behave in the desired way) – that are directly related to the way roads are designed or traffic is managed. These road users might have behaved in the desired manner if they had understood how to behave in a certain situation, were able to perform that behaviour and felt comfortable with it. The difference between willingness and disobedience is that the latter is more an intrinsic personal characteristic, independent of the traffic situation. This book is purely focused on the factors that are related to the way roads are designed or that traffic is managed and will deal with willingness.

This book was initiated by *Rijkswaterstaat* and is the result of several years of cooperation between *Rijkswaterstaat* and experts on human factors in the Netherlands. Together they have summarized the most important knowledge of human factors and road traffic, laid down in this book, into 'Ten Golden Rules'. These are also to be found as a brochure, widely used within *Rijkswaterstaat*, which has already received positive international attention.

We believe that anyone who is involved in road design or traffic management should have knowledge of the contents of this book and should also consider reading the brochure 'Ten Golden Rules' for a better understanding of the behaviour of road users.

Unfortunately, one of the experts involved in the preparation of this book passed away early in 2009. We therefore dedicate this book to Talib Rothengatter for his knowledge, dedication and his always critical reflection of the human mind.

Ton van den Brink,
Rijkswaterstaat, Delft

Michel Lambers,
Adviesdienst Mens en Veiligheid, Zevenhuizen

Chapter 1
Introduction

Maria Kuiken, Michel Lambers and Jan Theeuwes

Most of us will experience at least one car collision in our lives. Such a collision can be a minor fender bender or a large crash involving major injury and damage. The most common cause of car collisions is driver error. Often we may want to blame the driver who hit us, or possibly blame ourselves for not concentrating enough on the road. In other cases we may not realize that we were speeding or that we tried to negotiate a bend at too high a speed. The typical response to driver error is to blame the person who made the error. It may go on your driving record; your insurance premium may go up; and sometimes you are forced to go back to the driving school so they can teach you not to make mistakes in future. The bottom line is that most road authorities, including the police, believe that the person who made the error is responsible. The idea is that if you punish the person and tell them to behave better, the problem will go away.

Unfortunately, typically the problem does not go away. People still make mistakes, and in most cases they cannot even help making them. Drivers try to do their best but the task is complex and the environment is not designed to prevent errors occurring.

This book starts from a different perspective. We believe that drivers do not want to make errors. We believe that in many cases the design of the environment can be further adjusted to human capabilities. We argue that safety should be considered a system property to be built into the road system. The central theme of this book is how design principles can reduce the probability of errors while driving.

Many books focus on individual differences and how they relate to traffic safety, such as accident proneness, gender differences, age, alcohol and the effects of drugs. Others focus on vehicle safety factors such as airbags, anti-lock brakes, navigation systems, intelligent cruise control and other modern gadgets. Even though these topics are undoubtedly important for traffic safety, this book takes a unique approach as it focuses on the road environment itself. The book provides the background for those who want to know more about the effects of road design on driving behaviour. We use a systemic approach to allow a better understanding of why and in what circumstances drivers may commit errors. This understanding ultimately will lead to road systems that prevent (fatal) errors from happening. The book contains an overview of the current models and theories on human performance and human behaviour in traffic that are relevant for all those involved in designing safe road systems.

Background

Traffic Safety: Self-Explaining Roads and Sustainable Safety

The Dutch approach to traffic safety is the starting point. The reason for this is simple: the Netherlands maintains a high score on safety relative to other countries. In the last five years there were 4.6 casualties for every 100,000 inhabitants, which makes the Netherlands second in the European Union (EU) for road safety (Public Works and Water Management, Transport and Water Management Inspectorate 2006).

One of the building blocks of this success is the introduction and implementation of the so-called 'sustainable safety' approach. Sustainable safety specifies that safety should be a design requirement in road traffic systems. It advocates a proactive approach in which human characteristics are used as the starting point. These characteristics refer on the one hand to human physical vulnerability and on the other hand to human (cognitive) capabilities and limitations (Wegman and Aarts 2006). Roads should be categorized according to their function and as much as possible be 'self-explaining'.

Maximizing Throughput: Focus on Utilization

Safety is not the only issue for road authorities and policy makers in the Netherlands. On the motorways there has been a growth in (motorized) mobility that has resulted in a relatively dense road network. This has resulted in a comprehensive road construction programme which has been under way since the seventies. Despite this building programme and later additions (for example, road widening), the increase in the scale of the infrastructure has failed to keep pace with the rise in mobility. The conclusion has been reached that more infrastructure is not the way of the future. Building new roads not only involves high costs. There is also an increasing concern in society that the construction of new roads and/or the expansion of existing ones will affect the 'liveability' of the environment.

A new approach consists of making better use of the existing infrastructure. Optimal utilization can be defined as increasing road performance by optimum use of the physical space within certain parameters. These parameters could include pavement width or road alignment, but could also be legal in nature (Stembord, Van den Brink and Helleman 2002). This change of strategy did not only occur in the Netherlands, but can also be seen in most densely populated areas. The importance of a better utilization of existing infrastructure has now been incorporated in Dutch long-term plans for transport and infrastructure.

The constant struggle with traffic jams and the focus on 'utilization' led to a strong emphasis on traffic management. In situations where traffic is not meeting safety and throughput objectives,[1] action is taken to make the traffic process more

1 Recently, environmental objectives have been added to the safety and throughput objectives.

efficient. Numerous measures have already been implemented that try to ensure optimum utilization of the present road network infrastructure in the Netherlands. Some of these measures are designed to facilitate, with optimum use of the existing cross-section. Examples are the use of the hard shoulder in peak hours and dynamic/dual mode marking. Other measures are flexible and adjustable speed limits, and ramp metering. The success of traffic management depends on the measures taken and on the reaction of the road users. Unfortunately, the results have shown that the effects of particular measures did not always meet the expectations.

Technical Solutions

Rijkswaterstaat, the national road authority in the Netherlands, is responsible for the design and implementation of roadworks and employs many traffic engineers to that end. Problems within traffic management are usually considered from a technical point of view instead of considering the human factors (Lambers 2008). Consequently, traffic engineers primarily solve problems without consulting experts in human factors. Lambers describes the process as follows:

> When there is problem in managing traffic such as a capacity bottleneck, the first step is typically to take the problem out of its context and analyse it at an abstract level. When analysed, a particular option (for example capacity redistribution) is selected as a potential solution. The solution is usually practical and concrete. However, the solution developed from the engineering perspective. The measure is then implemented, but all too often the effects are disappointing due to unexpected and undesirable behavioural side effects on unforeseen consequences. Most engineers predict the behaviour of the road user on the basis of common sense. Sometimes it works, sometimes it doesn't. Sometimes the actual behaviour of the road users differs so much from the desired behaviour, that costly measures become counter effective or superfluous.

To further optimize traffic management measures and avoid failure, we all would like a situation where human factors knowledge can be taken into account at an early stage; or, as Barjonet (2001: ix) says:

> With a little more psychology, traffic would be improved considerably, in particular on the side of road insecurity which level and human cost remain, in many countries, unbearably high.

How do we realize this goal? In principle, traffic psychology can support engineering measures. Over the years many studies have been conducted by (traffic) psychologists and there is a large body of human factors knowledge available. In reality, the available human factors knowledge often does not reach the traffic engineer.

The Contribution of Traffic Psychology

The field of 'traffic psychology' can be defined as 'the study of the behaviour of road users and the psychological processes underlying that behaviour'. Psychology can contribute to the interdisciplinary traffic and transport sciences with its models, metaphors and methodologies that are specific to its disciplinary approach. It can determine and predict what effects the possibilities and constraints of the transport system will have on decision making processes of the actors. Conversely, it can determine and predict what demands the actors will pose on the transport system components (Rothengatter 2001). From this perspective it is clearly relevant that engineers are aware of the characteristics of the people for whom they are designing and managing the road and transport system (Fuller and Santos 2002). This is even more important considering that road user behaviour is the most important single collision-contributory factor.

Crashes are a negative and undesirable outcome of traffic. Because of the fact that traffic is a result of an interaction between humans, vehicles and roads – where the last two are relatively constant factors – most crashes are a direct result of human factors. It is estimated that major driver errors or violations are the main cause in about 80 per cent of crashes. However, this does not mean that road users are to blame for everything. Often, human error is a result of not taking human factors into account.

Ten Rules for Human Factors in Traffic

To provide road authorities and other governmental bodies with practical rules for road design, so-called 'golden rules' for human factors in road traffic have been developed. This has resulted in a list of ten basic principles that represent the main knowledge that should be taken into account in road and traffic management design. The rules address the following domains (see Figure 1.1):

- information elements on the road and roadside (for example, signs and markings);
- the interaction between different information elements (for example, no conflicting message between traffic signs and road markings);
- the situational context (for example, self-explaining roads and graceful degradation);
- traffic and the driving task;
- characteristics of road users as human beings (for example, the role of age, social aspects and communication).

This book presents the rules and their theoretical foundations. The road user and road user behaviour are considered from different theoretical perspectives. In view of the many determinants of road user behaviour it is not surprising that

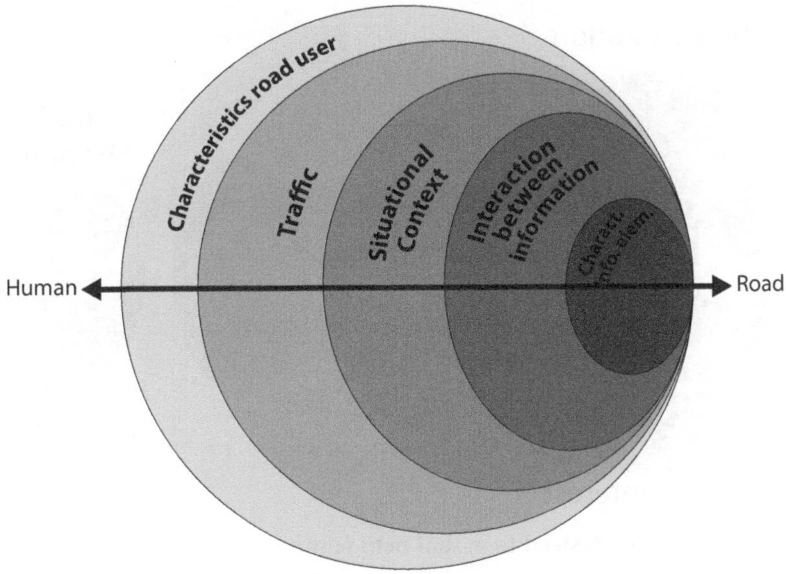

Figure 1.1 Structure of the rules: Five theoretical perspectives

there is no single theoretical framework for traffic psychology research. Some theories may be more applicable than others, depending on the problem. For instance driver behaviour can be studied in relation to visual search, workload, risk perception, attitudes and age, to name but a few factors that are thought to be determinants of road user behaviour. Which human factors must be taken into consideration when contemplating new measures or interventions? Many factors will influence the outcome (see Figure 1.2). First, drivers must *comprehend* a measure and understand what is expected of them. Second, they must be *able* to perform the desired behaviour. Third, the driver must be *willing* to perform the task. Behavioural options decrease due to limited comprehensibility of the measure, local alternatives and individual interests.

The motorway engineer has a major role to play in achieving safe mobility by bringing about substantial future road safety improvements. However, a 'human-centred design' requires that engineers are willing to learn about human limitations and motivations. This book provides the fundaments on which the most important rules of behaviour rest. The ten chapters of this book each address a specific rule.

The road and the traffic system should be designed according to the concept of Self-Explaining Roads (SERs). This concept advocates a traffic environment that elicits safe behaviour simply by its design.

The central theme of Chapter 2 is that road design should take into account human processing capabilities and build these into the road and vehicle environment.

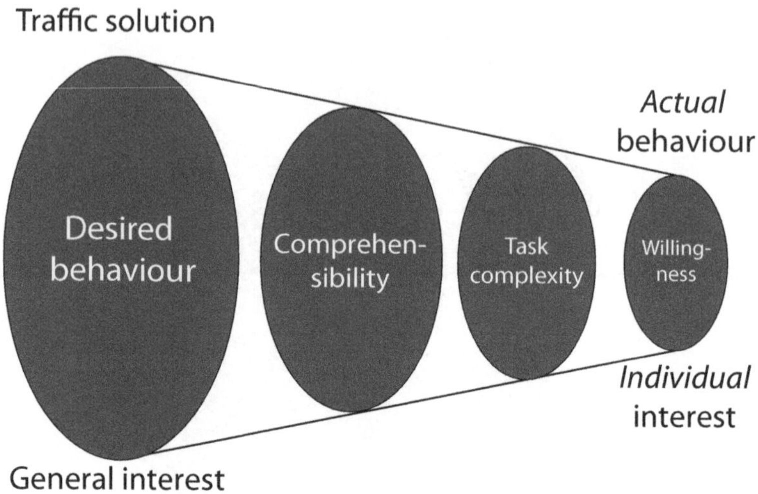

Figure 1.2 From desired to actual behaviour

Traffic systems which have self-explaining properties are designed in such a way that they are in line with the expectations of the road user. The chapter describes the main principles of SERs and gives a few good practice examples.

> *One should take into account inadvertent aberrations of the system (for example, a broken signalling device) or of the road user (is not always paying full attention to the driving task). Inadvertent aberrations should not result in accidents (principle of graceful degradation).*

The central issue addressed in Chapter 3 is that integrated systems design should anticipate the potential occurrence of inadvertent aberrations (for example, traffic signals breaking down, road users taking the wrong exit, missing a traffic sign etc.) and build in safety measures to ensure that even a breakdown of particular components does not immediately result in collisions (principle of graceful degradation). The road system should be robust in design and meet the sustainable safety principle of 'forgivingness' of the environment and of road users.

> *Driving involves three hierarchical task levels: viz. navigation, guidance and control. For each level task load varies depending on time and place. In road system design it is important to distinguish between all three levels.*

Driving is a complex task. Chapter 4 provides an overview of the main theories and models of driver behaviour and performance and their implications for road design.

When approaching task overload, road users will adapt their behaviour such that the load becomes lighter. Additional tasks will be ignored or skipped. When actual task overload occurs, rational behaviour may cease to exist and drivers may fall back to basic, rather reflexive behaviour tendencies.

The underlying idea of this rule is that people have a limited amount of processing resources available. If the demands of the task exceed the amount of processing resources available one speaks of task overload. With increasing demands for attention from in-car systems as well as from relevant and irrelevant information along the roads, it is important that workload will remain an important concern for traffic safety. Chapter 5 deals with the question how to design environments that do not induce a high workload for users.

Individual information carriers should meet basic ergonomic principles to be visible, clear and understandable for the road user.

Chapter 6 focuses on characteristics of individual information carriers. Road users must be able to perceive the roadway and understand what the motorway designer and traffic engineer intended to communicate to them. The chapter describes the way people perceive and collect relevant information from the environment and shows some implications for designing road systems.

Individual information elements should be mutually consistent and uniform within their context and should not confuse the driver.

Conflicting information or information other than expected may cause distraction, confusion, longer reaction times etc. Information carriers (location, timing and the content) should meet driver expectation. Consistency in design and application of uniform information may help drivers better predict what is to be expected and so anticipate what is coming instead of being surprised at the wrong moment.

When introducing particular measures it is likely that there will be adverse side effects. The driver is an integral part of the traffic system and will adapt its behaviour when necessary and possible. Any change within this system may lead to behaviour adaptation.

In Chapter 8, the underlying notion is that people do not simply ‘undergo’ changes in the environment; rather, they respond to changes in the environment by changing and adjusting their behaviour. From a safety point of view, this implies that expected benefits of particular measures may not be as large as originally expected. When introducing particular measures, it is important to consider the whole road system and which side effects are likely to occur.

The road user maximizes their behaviour to reach their goals and adapts their behaviour to that end. In this process are also included non-driving activities such as telephone conversations, entertainment, relaxation and work.

Drivers have their own goals and they will maximize their behaviour to reach them. It is shown that drivers match capability and task demand and try to maintain an acceptable task difficulty in all situations. They may, however, decide to deviate from normative behaviour or to perform other tasks while driving. Chapter 9 describes how, by incorporating a social psychological perspective, the chances of successful engineering will be increased.

Road users differ and human performance varies over time. Capabilities, social factors and momentary factors influence traffic behaviour. It may be useful to distinguish between groups. In the design of road and traffic management measures it is recommended to take characteristics of the target group into account.

Road users adapt not only to environmental factors but also to (momentary) fluctuations at an individual, personal level. Performance is vulnerable to a wide range of influences, such as: constitutional characteristics – experience; task-generated momentary factors (for example, emotion, fatigue); and non-task generated factors (such as aggression, stress, alcohol). Engineers have the challenging task of taking account of driver capabilities and individual as well as momentary differences in these capabilities.

Communication can contribute to behavioural change. The effectiveness of communication strategies depends on many factors.

In Chapter 11 different strategies and principles are described that can be used for behaviour-focused interventions. Road authorities communicate with road users through the basic design of the road (preferably self-explaining) and through billboards and signs with information or prompts. Theories of social influence can assist road authorities in making behavioural interventions more effective. Mass media campaigns are generally far less costly than individualized information. Individualized information tends to be more effective in changing behaviour because the information can be tailored to the needs and preferences of individuals being targeted.

The ten rules focus primarily on the driver interaction with the road environment. The central theme is the design of the road infrastructure, the readability of the road and the design of traffic management measures. The aim of this book is to bring together human factors and motorway engineers, and in this way contribute to the design process. Taking into account the determinants of road user behaviour will lead to new interventions (and design) and will increase the effectiveness of current measures and interventions. This book focuses primarily on the design of road systems. Relevant domains that have not been addressed are driver support and assistance systems (in-car and roadside technology) and mobility in a broader context. In the coming years new information and support systems hold the promise of great improvements both in the efficiency of transport systems and in the safety of road users. These systems have the potential to increase the quality and reliability of information available to drivers about their immediate driving environment.

References

Barjonet, P.E. (ed.) 2001. *Traffic Psychology Today*. London: Kluwer Academic Publishers.

Fuller, R. and Santos, J.A. 2002. *Human Factors for Highway Engineers*. Oxford: Elsevier.

Lambers, M.G.F. 2008. *Human Factors and Human Error as Part of the System: A Psychological Approach*. Paper presented at Transport Research Arena Europe, April 2008, Ljubljana.

Public Works and Water Management, Transport and Water Management Inspectorate, 2006. *Safety Balance*.

Rothengatter, J.A. 2001. Objectives, topics and methods. In: Barjonet, P.E. (ed.), *Traffic Psychology Today*. London: Kluwer Academic Publishers.

Stembord, H.L., Van den Brink, T.J.D and Helleman, D.E. 2002. *Dynamic Cross-Sections: Increased Capacity on Existing Infrastructure*. Transportation Research Board (TRB) Annual Meeting, Washington, DC, January 2002.

Wegman, F. and Aarts, L. 2006. *Advancing Sustainable Safety: National Road Safety Outlook for 2005–2020*. Leidschendam: SWOV Institute for Road Safety Research.

Chapter 2
Self-Explaining Roads and Traffic System

Jan Theeuwes

The road and the traffic system should be designed according to the concept of Self-Explaining Roads (SERs). This concept advocates a traffic environment that elicits safe behaviour simply by its design.

Theoretical Background

It is estimated that 90 per cent of traffic accidents are related to human error (for example, Treat et al. 1977). Because education and providing information only has a limited effect in changing people's behaviour (see Wakefield, Loken and Hornik 2010) it is crucial that the road environment is designed in such a way that it is adjusted to human processing capacity limitations. When designing a road, one should take the human processing capabilities into account and build these into the road and vehicle environment (Theeuwes and Godthelp 1992, 1993, 1995). Traffic systems which have self-explaining properties are designed in such a way that they are in line with the expectations of the road user. The so-called 'Self-Explaining Road' (SER) is a traffic environment which elicits safe behaviour simply by its design.

In order for road design measures to have the desired effect on driver behaviour, the road environment should be perceived by the driver in the way it was purposely designed. Note however that perception is an active construction process. It is the result of an interaction between sensory input, expectations and other information processing characteristics of the driver. Therefore, it is quite feasible that drivers may perceive the road environment in a manner different from the one the road designer initially had in mind. For example an urban four-lane road may be designed for a driving speed of 50 km/h; yet, if drivers perceive the road as 'a type of motorway', even though it is in a built-up area, it is likely that they will behave and treat the road as a motorway.

In general it has been estimated that over 90 per cent of the information that a driver has to process is visual (Hills 1980). Since the visual system is limited and the driving environment is relatively complex, in many circumstances perception of the road environment will rely on top-down expectations. In other words, drivers will perceive those events that are in line with their expectations and will overlook events that are not in line with their expectations. Typically, collisions occur because drivers do not expect particular events to happen and do not anticipate them adequately. Studies investigating errors occurring in actual traffic show that expectations play a crucial role in the occurrence of collisions. Accident data show that a large portion of drivers involved in crashes do not act too late; rather, they do not act at all to avoid the collision (Sussman et al. 1985).

Concepts and Schemata

As noted, perceiving our environment is an active process of selecting, organizing and interpreting the information sent to the brain by the senses. Information processing depends very much on how this information fits with our internal mental structures, called 'schemata'. The repeated experience of similar events and situations generates mental structures and rules that represent them. By developing these schemata, the mind creates order and structure from the mass of stimulation, finds regularities and comes to expect them in the future.

The idea that people categorize objects is based on the general notion that people try to structure their world. It allows us to understand individual cases that we have not seen before. From studies investigating knowledge representations, it is known that people classify objects as belonging to a particular category (for example, Rosch 1978). Through experience of particular objects, we develop internal representations that contain the typical characteristics of a particular category. The category to which an object belongs tells something about the characteristics of the object as well as the behaviours associated with it. Categorization is the grouping of experiences into categories (groups, classes). To categorize an object means that one considers it not only equivalent to other stimuli in the same category but also different from stimuli outside that category.

Subjective Categorization of Road Environments

Most research on categorization has concerned objects. In many respects categorization of environments shows the same phenomena as that of simple objects. The categorization of environments also shows a hierarchical structure with a basic level (for example, Russell and Ward 1982). In addition, people distinguish environments not only on the basis of physical characteristics but also on the basis of the goals or behaviours that take place in that environment. Finally, there are indications that a 'graded structure' is present within environmental categories: that is, one environment is a better example than another for a particular category. For example, classifying a road as a motorway will immediately instantiate particular expectations regarding the physical characteristics of the road (blue overhead signs, road markings, emergency lane and so on); the behaviour of other road users (fast driving, overtaking, no pedestrians); and the behaviour road users are allowed to display.

There are two general principles for categorization: cognitive economy and perceived world structure. Cognitive economy refers to the function of categorization in that it provides maximum information with the least cognitive effort. Perceived world structure refers to the structure of the information and asserts that the perceived world comes as structured information rather than as arbitrary or unpredictable

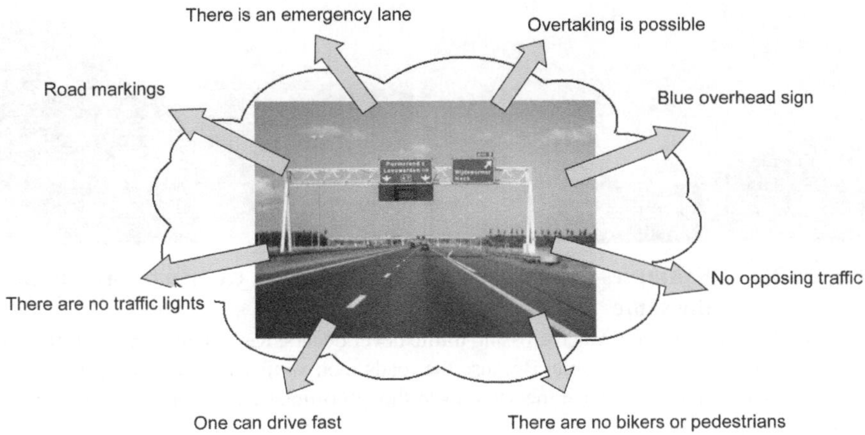

Figure 2.1 Classifying a road as a motorway activates schemata about what to expect on this type of road, how to behave and how other road users will behave

attributes. Thus maximum information with least cognitive effort is achieved when categories map onto the perceived world structure as closely as possible. Applying the principle of cognitive economy to the road environment entails that road users try to reduce the large number of roads that exist in the 'real' world to a few behaviourally and cognitively relevant road categories. It is to the road user's advantage to only differentiate between road categories when the difference is behaviourally relevant. To categorize a road as belonging to a certain category implies that it is similar to all other roads within that category and different from roads outside that category. For the road environment, the perceived world structure suggests that road users see the environment as consisting of a set of attributes that are highly correlated. The world does not consist of a set of attributes that are picked randomly. Thus, through experience with the road environment, road users develop a perceived world that contains attributes that are likely to occur in combination. In other words, if one sees a four-lane motorway one expects some road markings, an emergency lane and fast traffic that moves in the same direction (see Figure 2.1).

Previous studies investigating how people categorize road environments demonstrated that the subjective categorization (the way people categorize the roads in their 'heads') may not necessary match the official road categorization (Theeuwes and Diks 1995). It supports the idea that road behaviour is related to the appearance of the road and that road characteristics and traffic behaviour are cognitively integrated by drivers into subjective categories. Importantly, it seems that the road categories that are not categorized in clear clusters are also the types of road on which many collisions occur. Again it stresses the importance of adequate road design.

Figure 2.2 Examples of road types that road users categorize as basically the same

Note: Left, a motorway on which crossing traffic never occurs; Right, a motorway (100 km/h) where crossing traffic is possible. Because the roads look similar people categorize them as being the same and will behave the same even though officially these are different types of road.

Theoretical Model

Because behaviour is controlled by what people expect, subjective road categorization drives and controls traffic behaviour. Figure 2.3 presents a conceptual model of how road categorization affects a host of different expectations regarding one's own behaviour, the behaviour of other road users, road elements and type road transitions. For example, categorizing a road in a particular way will determine how fast one will drive (Theeuwes and Diks 1995; Kaptein and Claessens 1998) and how fast one expects other road users to drive. Also, dependent on the road type, one has particular expectations about what manoeuvres are allowed: for example, whether one can turn around, overtake, stop and so on. Also, depending on how one categorizes a road, one expects that particular road elements such as traffic lights, roundabouts or stops signs could be present. Finally, expectations regarding transitions play an important role. For example, if one thinks one is driving on a motorway, one expects that transitions to other roads will be relatively infrequent; and if there is a transition one will leave the motorway to enter the next level of road, for example a provincial road.

In a driving simulator study, Kaptein and Claessens (1998) investigated three different types of road category outside the built-up area. One type was basically the standard method of road design; one design was what was labelled as a self-explaining road (SER); and one was a combination of standard and self-explaining design. The SER design showed a strong overlap in the type of properties within one road category and hardly any overlap between the road categories. Kaptein and Claessens showed that road users were better able to classify the SER design as belonging to one or the other road type. For the basic standard and mixed designed there was much less consistency in classifying these roads. The most important finding from this study was that when subjects were exposed to these different types of road in the driving simulator they chose more consistent and

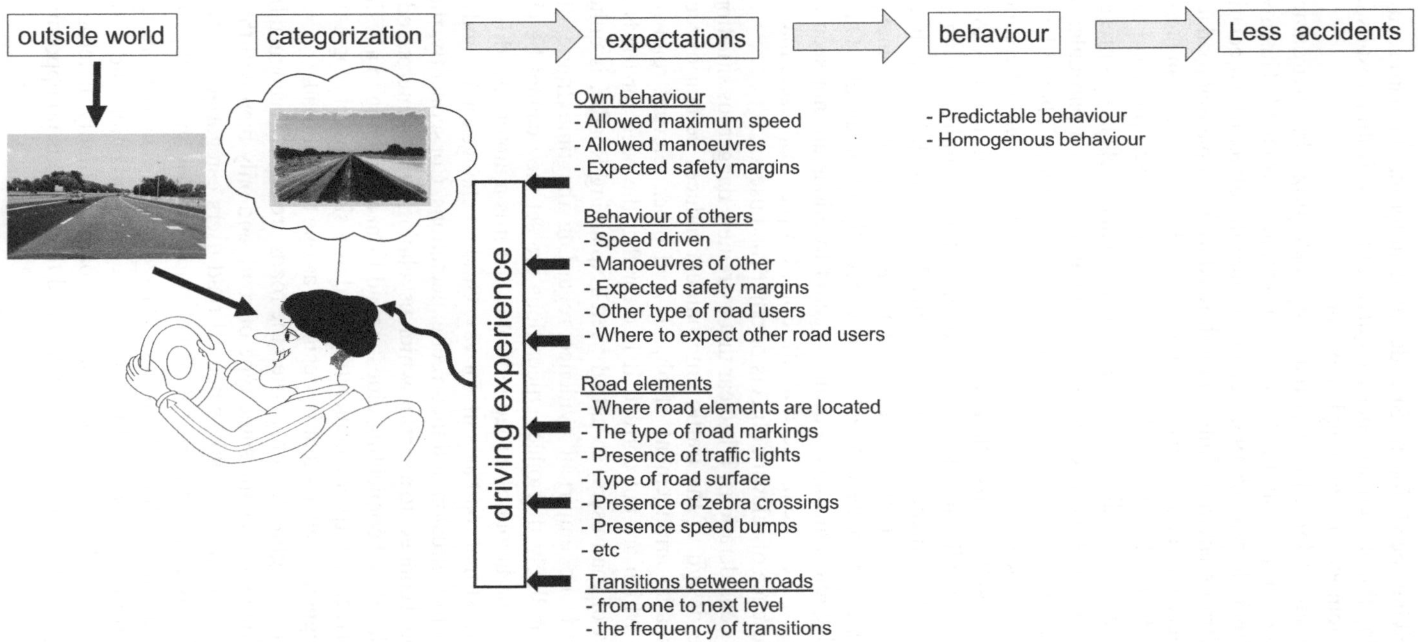

Figure 2.3 **A model explaining how subjective road categorization affects road behaviour, and how this feeds back to the road user**

homogenous driving speeds for the SER design than for the standard regular road design, indicating that systematic and unequivocal categorization results in more systematic and homogenous driving behaviour.

It should be stressed that the way road users categorize the environment also has a major effect on the way the traffic environment is scanned. Theeuwes and Hagenzieker (1993; Theeuwes 1996) demonstrated the effect of expectations on visual search in traffic environments. They showed for example that road users are less likely to detect and see a cyclist or a traffic sign when they are in an unexpected location within the road scene. To illustrate this with an example, road users are less likely to see a cyclist riding on the main road when a cycle lane is present. The underlying notion is that each traffic environment activates a particular schema which induces expectations of where particular target objects should be. Search behaviour will be biased towards those portions of the visual field where the target is expected. It should be realized that the effects of contextual-driven search might be very strong in conditions of high workload – for example, driving in busy traffic in urban environments – or under reduced sight conditions such as when driving in the dark or in twilight. In these situations, rapid resource inexpensive and conceptually driven feature detection are especially advantageous. Our notion is that potentially relevant objects at unexpected locations are not seen too late but, in most cases, they are not seen at all. For example, road users may simply overlook the presence of relevant objects or other road users.

Given these considerations, it is clear that extremely dangerous situations may occur when the design of the traffic environment induces certain expectations regarding the spatial arrangement of objects and other relevant road users in that road scene which are not correct. It is clear that the need for roads that fit the expectations of road users, so-called 'Self-Explaining Roads', is quite high. Although there is no empirical evidence regarding the time frame of these expectations, it is likely that once expectations are set they cannot be changed easily. If initially the layout of a road indicates that it is a motorway, drivers will continue to interpret that road as a motorway. Gradual changes into another type of road will lead to confusion and incorrect expectations. Consequently, the same type of road should connect to a section which psychologically is interpreted as one unit: for example, a road connecting cities; a road from a shopping to a residential area. Because drivers interpret a connecting road as a single psychological 'unit', it should be designed as such – that is, designed as one type of road.

As is clear from Figure 2.3, only when the roads are clearly recognizable will all users categorize them as the same type of road, and only then will this result in homogenous expectations. Only when all road users categorize a road in the same consistent way can one expect their behaviour shown on this type of road to be homogenous. It is important to note that when expectations are met this will feed back to the road user and reinforce the categorization and associated expectations. If the physical appearance of roads is highly heterogeneous, consistent categorization will not take place, and then there is no opportunity for road users to learn anything. Road user behaviour displayed on these roads will not be consistent and errors are likely to occur.

How to Design According to SER Principles

As outlined above, traffic systems should be designed so that they are self-explaining. A Self-Explaining Road (SER) is a traffic environment which elicits safe and consistent behaviour among road users simply by its design. As noted, it is important that road users are able to categorize a road immediately and consistently as belonging to one or other road type. In the Netherlands the notion of Self-Explaining Roads resulted in the redefinition of three road types: flow, distributor and access road. Their essential characteristics are shown in Figure 2.4. The three general principles of SER design state that roads should be:

- *Easily recognizable*: Roads that have the same function, the same speed profile and the same mix of road users should look similar.
- *Easily distinguishable*: Roads of different categories should look different. In other words, the layout between the different categories should be different.
- *Easily interpreted*: Desired driver behaviour should be clear and the characteristics to differentiate the different road categories should induce this behaviour.

Essential Recognition Characteristics	Flow-road		Distributor road	Access road
Speed limit	120 km/h	100 km/h	80 km/h	60 km/h
Road sign	 (ASW, G01)	 (AW, G03)	no	 (A01(60) / E10)
Road side/edge marking	unbroken	unbroken	broken	broken or none
Separation of driving directions	wide median or barrier	double axis marking with 'green' filling, or median	double axis marking (without filling), or median	no

Figure 2.4 Essential road characteristics for three well-defined road categories as employed in the Netherlands: Flow, distributor and access roads (SWOV 2007)

Given these general guidelines any of the following principles may apply (from Theeuwes 2001):

- Unique road elements (homogeneous within one category and different from all other categories).
- Unique behaviour for a specific category (homogeneous within one category and different from all other categories).
- Unique behaviour should be linked to unique road elements (for example, *woonerfs*: obstacles, slow driving; motorway: smooth surface, fast driving).
- The layout of crossings, road sections and curves should be linked uniquely to the particular road category (for example, a crossing on a motorway should physically and behaviourally be completely different from a crossing on a rural road).
- One should choose road categories which are behaviourally relevant.
- The same road category should connect a section which psychologically is interpreted as a single unit (for example, a road connecting two cities).
- There should be no fast transitions going from one road category to the next.
- When there is a transition in road category, the change should be marked clearly (for example by rumble strips).
- When teaching the different road categories, one should include not only the name but also the behaviour required for that type of road.
- Category-defining properties should be visible at night.
- Road design should expel speed differences and differences in direction of movement.
- Road elements, marking and signing should fulfil standard visibility criteria.

Recent Studies Investigating the Effects of SER Design

In a recent study Weller et al. (2008) used the concept of Self-Explaining Roads to investigate the subjective categorization of rural roads in Germany. As noted, rural roads are particularly important because fatalities on them are about 60 per cent, compared to 10 per cent on motorways. Clearly, rural roads have different functions, large variations in speed limits and are unforgiving (that is, an error may be fatal). Often the geometry of the road is not designed intentionally but rather is usually the result of some historical development (for example, a road between two villages). Weller et al. developed a driver and driving behaviour model for rural roads which has a strong resemblance to the ideas underlying SER (see Figure 2.5). This model uses terms such as 'affordance and cues' (cf. Gibson 1968), which are the 'category defining properties' of the SER concept developed by Theeuwes (2000). As noted, these categories must be internally consistent, mutually exclusive and clearly distinguishable. Consistent with SER, the model also stresses that subjective road categorization is more important than objective 'official' categorization.

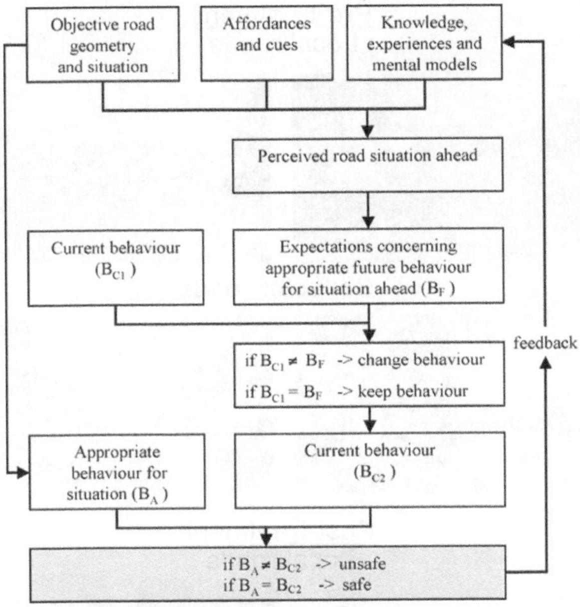

Figure 2.5 A driver and driving behaviour model for rural roads (as developed by Weller et al. 2008)

Weller et al. took 25 pictures of existing rural roads and had subjects categorize these pictures. By means of a cluster analysis (see, for example, Theeuwes and Diks 1995) three different road clusters were identified and labelled as 'demand', 'comfort' and 'monotony'. Importantly, as in Theeuwes and Diks, the subjectively estimated appropriate speed in the situation seen in the picture fitted well with the subjective categorization, implying that categorization is related to the type of behaviour users would show on these roads. This study confirms earlier findings in showing the relevance of subjective road categories for traffic behaviour (for example, Kaptein, Janssen and Claessens 1998; Riemersma 1988; Theeuwes 1998).

Several recent field studies have applied the principles of SER and tested whether changes in road environment would lead to adequate behaviour adaptation. A study by Charlton et al. (2010) carried out in New Zealand was designed to identify and develop guidelines that would enable the development of speed management approaches similar to self-explaining and sustainably safe approaches as developed in the Netherlands. The approach was formulated by the New Zealand Ministry of Transport (2004) as follows:

> The emphasis is not just on speed limit enforcement, it includes perceptual measures that influence the speed that a driver feels is appropriate for the section of road upon which they are driving – in effect the 'self-explaining road'.

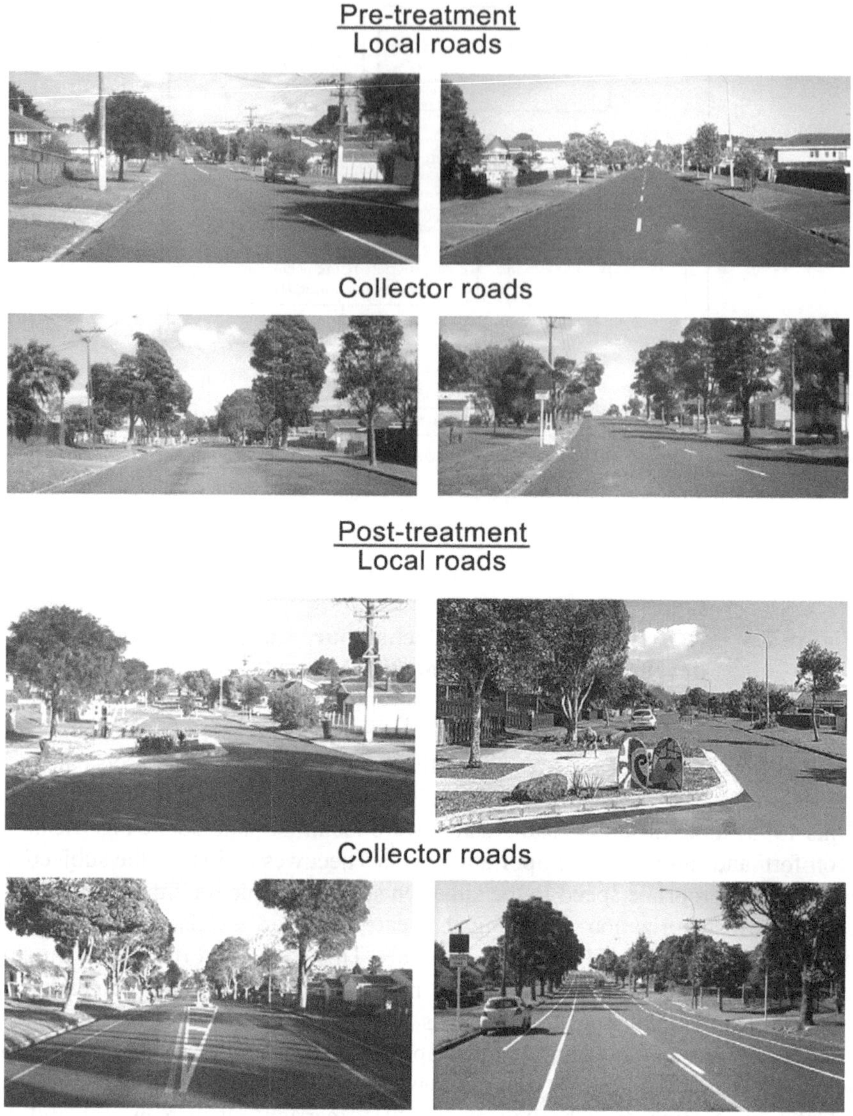

Figure 2.6 Existing roads in New Zealand (top) were redesigned by a Self-Explaining Road treatment (bottom) (from Charlton et al. 2010)

In this field study, particular roads received a SER treatment such that there were maximum differences between the different road categories. These SER adapted roads were compared to matched control roads. The SER design for local roads entailed landscaping and the creation of community islands to limit forward visibility, and the removal of road markings to create a visually distinct road

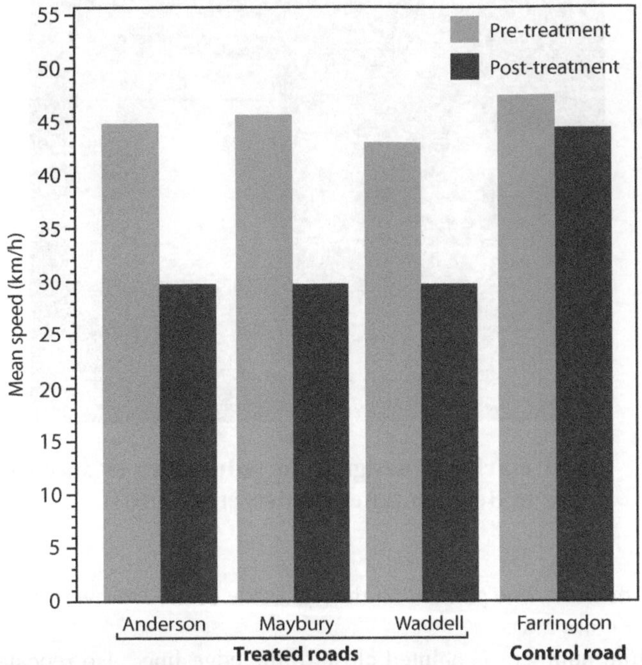

Figure 2.7 Effects of SER treatment on the speed driven on local roads in New Zealand (Charlton et al. 2010)

environment. Roads that were categorized as 'collectors' (for example, in our SER labelling 'distributor roads' – see Figure 2.4) had increased delineation, the addition of cycle lanes and specific design solutions for pedestrians. Figure 2.6 gives some examples of these roads. After the implementation of these measures speed data were collected for three months. The results were quite dramatic, as shown by a significant reduction in speeds driven on local roads and increased homogeneity of speeds on both local and collector roads.

Figure 2.7 presents the effects of the speed driven on local roads before and after the SER treatment. Overall, the New Zealand project was very successful in showing that the SER approach works. The project showed that by redesigning the roads a clear multi-level road hierarchy was established, with each level having a distinct 'look and feel' and discriminability of different speed profiles. Note also that residents' ratings of the appearance of their roads were significantly more positive following installation of the SER treatments.

Daniels et al. (2010) tested in both a field and in a simulator study whether additional road markings would reduce driver speed on 70 km/h roads. The researchers developed two types of road markings which complied with the United Nations Economic Commission for Europe (UNECE) European Agreement on Road Markings (1957):

Figure 2.8 Road markings designed to reduce driver speed on 70 km/h roads in Belgium (after Daniels et al. 2010)

- a 0.5m white line painted on the right side of the roadway every 50 metres (see Figure 2.8);
- a white number '7' painted close to the edge line, also repeated every 50 metres.

The results were somewhat disappointing. The field study showed the road markings had no effect on speed driven. The simulator study also showed no effect on speed, although there was some effect on lateral position.

The authors of this study indicated that the null effect was possibly due to the fact that the road marking were not conspicuous enough to be noticed. In the driving simulator, drivers were asked whether they noticed anything special. More than half did not notice the road markings – and one-third did not notice them even after they were explicitly told about the additional markings. Given the low conspicuity of these road markings, in an environment where nothing else changes it may not be surprising that such subtle changes go unnoticed.

Overall, worldwide there have been various attempts to redesign roads according SER principles. The most dramatic effects were seen in the study in New Zealand. This study illustrates that it is important not just to adapt one minor aspect of road design (road markings only, as was done in the Belgium study) but that a dramatic redesign, including landscaping, is necessary to change road categorizations and the associated user behaviour displayed on these roads.

Good Practice Examples

Since the introduction of its name and underlying psychological principles (for example, Theeuwes and Godthelp 1992, 1995) the concept of Self-Explaining

Roads is now one of the leading design principles in most European countries (see European Road Assessment Programme – EuroRAP). It was also recognized as one of the leading principles by the World Health Organization (2004), and non-EU countries have also adopted it – see Australia and New Zealand (Breen 2004) and the US (Campbell, Richard and Graham 2008).

Figures 2.9 and 2.10 show some examples from the Netherlands and the UK of suggestions for redesigns of roads according to SER principles. Looking at Figure 2.10 specifically, in the top picture, the road was narrowed both physically and visually using red-brick surfacing at the edge and in the distance, and the centre white

Figure 2.9 Examples of flow, distributor and access roads in the Netherlands (from SWOV 2007)

Figure 2.10 Examples of redesigns of roads in the UK according to SER principles

line was removed. Removal of the centre line suggests to the driver the possibility of meeting other vehicles head-on. In the middle picture the road was also narrowed and the centre line removed. In the bottom picture the transition from one type of road to one that is part of a village is indicated by narrowing, the removal of centre lines and the use of additional signing (from Kennedy et al. 2005).

Summary and Conclusions

The concept of Self-Explaining Roads is firmly rooted in the theoretical framework of subjective categorization and associated expectations (see Figure

2.3). Worldwide this concept has been embraced by many road authorities as an approach for redesigning the road environment. In the Netherlands this has resulted in a completely new approach, also known as 'sustainable safety', and a new categorization of the various types of road, known as flow, distributor and access roads. In other countries similar approaches have also been initiated. The most dramatic results were reported in New Zealand, were a substantial speed reduction was seen after redesigning the roads. In order to see an effect one cannot change just a minor aspect of the road environment; instead one needs a dramatic redesign, including landscaping, in order to change the road categorization and the associated traffic behaviour displayed on these roads.

References

Aarts, L.T., Davidse, D.J., Louwerse, W.J.R., Mesken, J. and Brouwer, R.F.T. 2005. *Herkenbare vormgeving en voorspelbaar Gedrag*. SWOV Report R-2005-17 (in Dutch).

Breen, J. 2004. *Review of the Road Safety to 2010 Strategy*. New Zealand Transport.

Campbell, J.L., Richard, C.M. and Graham, J. 2008. *Factors Guidelines for Road Systems*. Transportation Research Board, National Cooperative Highway Research Program (NCHRP) Report 600A.

Charlton, S.G., Mackie, H.W., Baas, P.H., Hay, K., Menezes, M. and Dixon, C. 2010. Using endemic road features to create self-explaining roads and reduce vehicle speeds. *Accident Analysis and Prevention*, 42(6): 1989–98.

Daniels, S., Vanrie, J., Dreesen, A. and Brijs, T. (2010). Additional road markings as an indication of speed limits: Results of a field experiment and a driving simulator study. *Accident Analysis and Prevention*, 42(3): 953–60.

European Road Assessment Programme (EuroRAP). Available at: http://www.eurorap.org

Gibson, J.J. 1986. *The Ecological Approach to Visual Perception*. Hillsdale, NJ: Erlbaum.

Hills, B.L. 1980. Vision, visibility, and perception in driving. *Perception*, 9: 183–216.

Kaptein, N.A. and Claessens, M. 1998. *Effects of Cognitive Road Classification on Driving Behaviour: A Driving Simulator Study* (Report TM-98-C048). Soesterberg: TNO Human Factors Research Institute.

Kaptein, N.A., Janssen, W. and Claessens, M. 2002. A study of subjective road categorization and driving behaviour. In: R. Fuller and J.A. Santos (eds), *Human Factors for Highway Engineers*. Amsterdam: Pergamon, 59–76.

Kennedy, J., Gorell, R., Crinson, L.,Wheeler, A. and El, M. 2005. *'Psychological' Traffic Calming* (TRL Report 584). Crowthorne: Transport Research Laboratory.

New Zealand Ministry of Transport. 2004. *The Current Approach to Speed Management*. Speed Management Stakeholders Seminar. Wellington, 17 November 2004.

Peden, M., Scurfield, R., Sleet, D., Mohan, D., Hyder, A., Jarawan, E. and Mathers, C. 2004. World Health Organization. 2004. *World Report on Road Traffic Injury Prevention*. Geneva: World Health Organization.

Rosch, E. 1978. Principles of categorization. In: E. Rosch and B.B. Lloyd (eds), *Cognition and Categorization*. Hillsdale, NJ: Erlbaum.

Russell, J.A. and Ward, L.M. 1982. Environmental psychology. *Annual Review of Psychology*, 33: 651–88.

Sussman, E.D., Bishop, H., Madnick, B. and Walter, R. 1985. Driver inattention and motorway safety. *Transportation Research Record*, 1047: 40–48.

SWOV. 2007. Factsheet *Herkenbare vormgeving van wegen*. Leidschendam: SWOV Institute for Road Safety Research (in Dutch).

Riemersma, J.B.J. 1988. An empirical study of subjective road categorization. *Ergonomics*, 31(4): 621–30.

Theeuwes, J. 1996. Visual search at intersections: An eye-movement analysis. In: A. Gale (ed.), *Vision in Vehicles V*. Amsterdam: North Holland.

Theeuwes, J. 1998. Self-explaining roads: Subjective categorization of road environments. In: A. Gale (ed.), *Vision in Vehicles VI*. Amsterdam: North Holland.

Theeuwes, J. 2000. Commentary on Rasanen and Summala 'Car drivers' adjustments to cyclists at roundabouts. *Transportation Human Factors*, 2(1): 19–22.

Theeuwes, J. 2001. The effects of road design on driving. In: P. Barjonet (ed.), *Traffic Psychology Today*. Amsterdam: Elsevier.

Theeuwes, J. and Diks, G. 1995. *Subjective Road Categorization and Speed Choice* (TNO-TM 1995 B-16). Soesterberg: TNO Human Factors Research Institute.

Theeuwes, J. and Godthelp, H. 1992. *Self-Explaining Roads* (IZF 1992 C-8, in Dutch). Soesterberg: TNO Human Factors Research Institute.

Theeuwes, J. and Godthelp, H. 1993. Self-explaining roads. In: J.L. de Kroes and J.A. Stoop (eds), *Safety of Transportation*. Delft: Delft University Press, 56–66.

Theeuwes, J. and Godthelp, H. 1995. Self-explaining roads. *Safety Science*, 19: 217–25.

Theeuwes, J. and Hagenzieker, M.P. 1993. Visual search of traffic scenes: On the effect of location expectations. In: A. Gale (ed.), *Vision in Vehicles IV*. Amsterdam: North Holland, 149–58.

Treat, J.R., Tumbas, N.S., McDonald, S.T., Shinar, D., Hume, R.D., Mayer, R.E., Stanisfer, R.L. and Castellan, N.J. 1977. *Tri-Level Study of the Causes of Traffic Accidents*. Report No. DOT-HS-034-3-535-77 (TAC).

Wakefield, M.A., Loken, B. and Hornik, R.C. 2010. Use of mass media campaigns to change health behaviour. *The Lancet*, 376: 1261–71.

Weller, G., Schlag, B., Friedel, T. and Rammin, C. 2008. Behaviourally relevant road categorisation: A step towards self-explaining rural roads. *Accident Analysis and Prevention*, 40(4): 1581–8.

Chapter 3

Resilience to Failure and Breakdown

Richard van der Horst

One should take into account inadvertent aberrations of the system (for example a broken signalling device) or of the road user (is not always paying full attention to the driving task). Inadvertent aberrations should not result in accidents (principle of graceful degradation).

Theoretical Background

People make errors and technical components may also display failures. Integrated systems design should anticipate potential occurrences of inadvertent aberrations (traffic signals breaking down, road users taking the wrong exit or missing a traffic sign etc.) and build in safety measures to ensure that even breakdown of certain components does not result directly in crashes (principle of graceful degradation). In case of technical system failure, the road user should still be able to cope with the situation in one way or another; and in case of human failure the road system should be robust in design and meet the sustainable safety principle of forgivingness of the environment and/or of road users. In this sense, it comes close to the definition of resilience by Hale and Heijer (2006), reflecting not only the flexibility of a system/organization to cope with unexpected and unplanned situations and respond rapidly to events but also the system's ability to anticipate and circumvent threats.

Human error is to be considered a major contributing factor to the occurrence of crashes. Even if people are highly motivated to behave safely, they will make errors that may result in crashes. As Wegman and Aarts (2006) state, current road traffic has to be considered as inherently dangerous since it is characterized by many latent errors. Figure 3.1 illustrates well how a chain of latent errors, together with dangerous actions, may result in a crash when it all fits, so to say (the arrow does not encounter resistance in one of the layers). It also illustrates that, although the occurrence of a chain with all the contributing factors 'in line' may be very rare, the likelihood of a given event is inherently present and, indeed, may occur in practice (Murphy's law).

Of course, a human-centred design and an integrated road traffic system approach taking human capabilities and limitations into account may reduce the occurrence of human error – but completely excluding them is utopian. Reason (1990) distinguishes different types of errors people make, dependent on the level of task performance – skill-based, rule-based, knowledge-based (Rasmussen 1985; for the levels of task performance, see also Chapter 4) – see Figure 3.2.

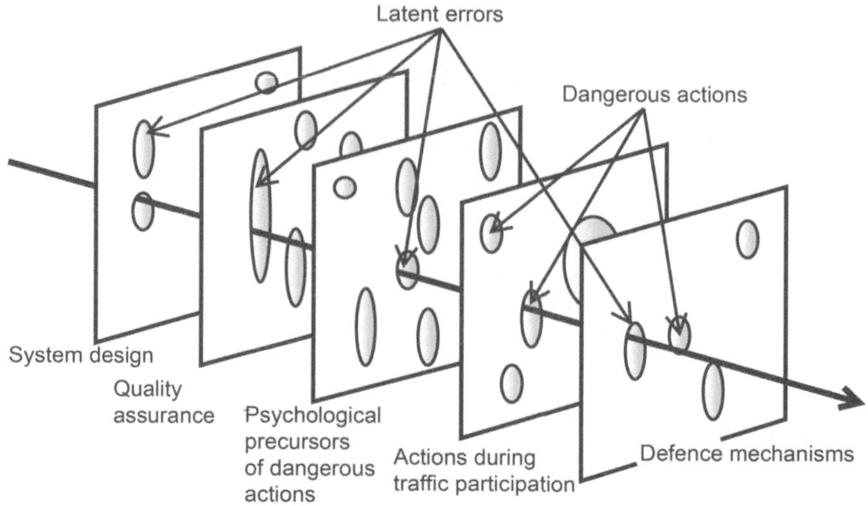

Figure 3.1 Schematic representation of the development of a crash (bold arrow) caused by latent errors and dangerous actions
Source: From Wegman and Aarts 2006; after Reason 1990.

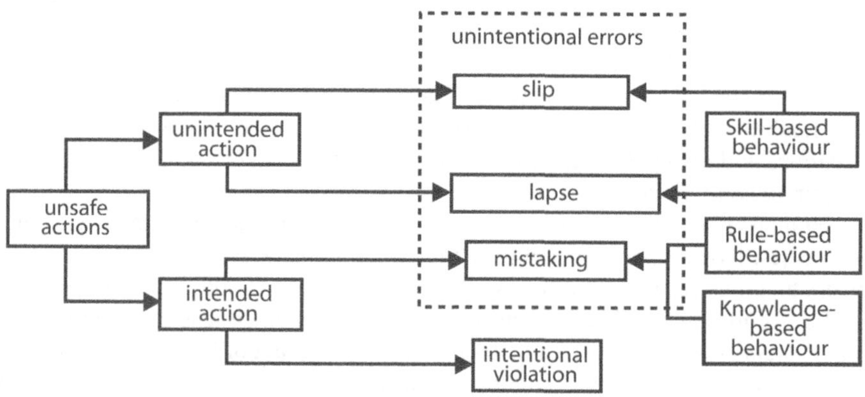

Figure 3.2 Taxonomy of dangerous actions
Source: From Wegman and Aarts 2006; after Reason 1990.

Roughly speaking, unsafe actions can be distinguished in unintended and intended actions. The latter may result either in *intentional violations* (the road user is aware that their behaviour is not allowed or may be unsafe but for some reason or another they consciously take the risk); or in *mistakes* (unintentional errors based on a wrong diagnosis or decision for the situation). For unintended actions the road user thinks they are doing the right thing and acting correctly but makes *slips* (the intention or plan is correct but the execution fails) or *lapses* (a failure to perform an intended action or forgetting the next action in a sequence) at the skill-based (automatic) level of task performance. Reason, Parker and Lawton (1998) extended different error types and behaviour in greater detail. For example, they made the distinction between compliant and non-compliant behaviour, the latter in violations and circumventions of safe operating procedures. They emphasize the importance of a system's intrinsic resistance to its operational hazards and propose a classification of rule-related behaviour as a means of understanding how rule quality, the potential outcomes of behaviour and the motives of individuals interact. For an organization, Reason et al. (1998) identify four distinctions:

a. between psychologically rewarding and unrewarding behaviour;
b. between violations and compliant behaviour;
c. between correct and incorrect actions;
d. between 'good' and 'bad' rules.

This results in ten types of rule-based behaviour that also apply to a great extent to rule-based behaviour of road users in the road system. These include:

1. mistakes;
2. correct improvisation;
3. correct but unrewarding violations;
4. correct violations;
5. mispliance;
6. incorrect but rewarding compliance;
7. misvention (mistaken circumvention);
8. incorrect but rewarding violation;
9. correct but unrewarding compliance;
10. correct compliance.

The limited available time frame of erroneous actions or the lack of action at the skill-based task performance level (mostly at the guidance and control level of the driving task, see Chapter 4) implies that these errors may rapidly result in safety-critical situations, whereas the consequence of making errors at the navigation level may result in lost or confused drivers without direct safety consequences (Alexander and Lunenfeld 1986). From a Norwegian in-depth accident analysis methodological study – *Driving Reliability and Error Analysis Method* (DREAM), by Sagberg (2007) – it appeared that a frequently occurring error mode was 'wrong direction'. An example with serious consequences is the mistake a driver may make when entering a motorway by turning into the exit lane. Inadequate road design through a mismatch between information at the route guidance level and the manoeuvring level of the driving task (see Chapter 4) may be an underlying factor; but overload or stress through a complicated route-choice

Figure 3.3 Dutch example of giving 'wrong-way' drivers additional information that they have made a mistake and should turn around

decision may also have made the driver act erroneously in their path choice. In addition, once such a decision is made and the action of turning has taken place it may be very difficult to make the driver aware of the mistake. In order to try to raise awareness of joining a motorway via the wrong entrance, in the Netherlands a set of directional arrows is painted on the roadway surface at exit lanes, together with a conspicuous 'entry forbidden' sign with the text '*Ga Terug*' ('Go Back') at the same location (see Figure 3.3).

Effects on Driving Behaviour

As people make errors, it is important that road systems design (the interaction between roadway design, traffic management, traffic rules, road users and vehicles) includes a certain amount of *forgivingness* – also one of the new pillars of the Sustainable Safety Principles in the Netherlands (Wegman and Aarts 2006). Redundancy can be an important element in preventing people missing relevant information or enabling them to adjust their decision making according to the situation at hand. Making the system 'forgiving' helps to reduce the consequences of human error as much as possible. Or, as Salmon, Regan and Johnston (2006) state, a road transport system should be made 'error tolerant'. A system should inherently have resilience against human error.

But not only human error should be considered. More technical components of the system may fail and the design should also be such that system failures are recognized in time by the road user so that with the information available they are still able to act appropriately given the conditions (the principle of 'graceful degradation'). Lack of information (especially when expected) or, even worse, contradictory information from different sources may confuse the road user and result in hesitation and unsafe behaviour. The important role of expectancy has already been detailed in Chapter 2. A 'dynamic signal off' also gives information to road users. For example, when the matrix signs of the Automatic Incident Detection System on motorways are off, drivers who are used to such a system do not expect traffic queuing in front of them. At signalized intersections, controller failure apparently occurs now and then (or, more regularly, the controller may be switched off late at night). In these cases, there should be a backup system to make it clear to drivers that the signal is not active (flashing yellow), and what type of behaviour is to be expected from them: for example, yielding to traffic from the left and right, clearly indicated by yield signs, yaw-teeth markings on the road surface etc.

Another example of drivers' part awareness and possible misuse of a situation occurs with shoulder lane use during rush hours, with a dynamic signalling system of green arrows when the rush-hour lane is active. Such a measure is in common use nowadays in the Netherlands on motorway sections that lack sufficient rush-hour capacity. But sometimes, when for traffic efficiency reasons it does not help to open the rush-hour lane (for example, due to a bottleneck downstream), traffic operators decide to keep the lane closed during the rush-hour. As some drivers know about the use of the emergency lane during rush hours, they may conclude that the system is not switched on or has broken down. They then decide to (mis)use the emergency lane as a traffic lane – according to Reason et al. (1998), 'incorrect but rewarding violation'. The initial idea to just switch off the matrix sign above the hard shoulder when it was not in use as a traffic lane had to be changed to an active signal of a red cross (see Figure 3.4). This was because road users during rush hour periods thought the system had broken down, the lane should be open and that they should be allowed to use the lane for driving.

Nowadays, the use of in-car navigation systems has become more and more popular, especially for drivers who experience difficulty finding their way in a complex and dense road network. But high-mileage motorists also frequently use in-car navigation systems for unknown destinations. In 2007 a study on the safety effects of in-car navigation systems indicated that 28 per cent of motorists owned a navigation system. Of these, 35 per cent used their system in 20 per cent of trips, whereas 15 per cent used their system in 80 per cent of trips (Vonk et al. 2007). As people rely more on such systems, the more vulnerable they become to possible breakdowns or temporary malfunctions, for example, due to global positioning system (GPS) signal failure. Drivers then need a fall-back system such

Figure 3.4 Emergency lane closed as a rush-hour lane by means of a red cross

as a paper map and orientation features in the environment to establish where they are and where they are heading. Therefore, appropriate roadside route signage is still needed for users of in-car navigation systems as well as for non-users.

How to Make the Road System More Resilient

Roundabouts

Compared to traditional priority intersections, roundabouts appear to be a sustainably safe solution, with an average reduction in accident occurrence of between 30 and 50 per cent (Elvik 2003). Apart from automatically inducing considerably lower approach speeds and the number of potential conflict points, a roundabout is also a good example of an intersection layout that allows road users to recover easily from an error in route choice or lane choice by staying on the roundabout and correcting their decision on the right exit. A roundabout is also a safe place to make a U-turn: in daily traffic drivers can simply go back the way they came, for example after dropping someone off at a railway station.

A rather new type of roundabout – the so-called 'turbo roundabout' – has physical separations between the lanes for different directions on the roundabout itself. This design has resulted from studies that clearly illustrate a higher capacity and better throughput of traffic, with less waiting time for individual road users (Fortuijn and Carton 2008; Fortuijn 2009). As a consequence of the physical lane

Figure 3.5 Example of a turbo roundabout in the Netherlands with physical separation between lanes so no correction to direction or lane can be made once a driver has entered the roundabout

dividers, drivers have to select their route before entering the roundabout because changes in directional decisions can no longer be made while on the roundabout (see Figure 3.5). In spite of the higher traffic efficiency, road users unfamiliar with this layout definitely do not like this type of roundabout as it gives them no opportunity to correct an erroneous route choice or lane choice once they have entered the roundabout. Therefore, this type of roundabout does not allow any resilience against human error or indecisive behaviour.

A behavioural observation study at two high-volume signalized turbo roundabouts indicated that, on average, 15 error manoeuvres per hour occurred there (van der Horst, Martens and Kik 2008). At both roundabouts the left-turn manoeuvre appeared the most problematic. Turning left directly beyond the first traffic light was considered to be the most serious problem because one enters the roundabout against the driving direction ('ghost riding'). While on the roundabout, the path choice for traffic turning left at the second traffic signal stands out most prominently. Complex error manoeuvres (such as switching carriageway several times, resulting in at least one complete lap) indicate that drivers do not know where to go or have lost the right path choice completely. From survey results it can be concluded that the interviewed road users are reasonably positive about signalized turbo roundabouts, but they indicate that habituation is very important. Many road users admitted that they made several mistakes (such as wrong lane choice) when the situation was new. Many respondents also indicated that they frequently see other road users making similar mistakes.

Redundancy in Design

People order their world in a few logical categories (see Chapter 2). In daily life, this is very handy as so much overload in information is presented, and structuring our world is a means to cope efficiently with this. Everything that functionally looks similar is put in the same category, whereas elements that are considered different in use are categorized differently. These principles can also be applied to designing better roads. Within one type of road everything has to be the same or similar as far as possible, whereas among different road categories differences should be as large as possible. Several road categories that are distinguished by relatively small differences may have a function for the road designer and administrator, but they may be of little use to the road user (see Chapter 2). For example, research by Kaptein, van Hattum and van der Horst (1998) reveals that if people have to learn to distinguish new, unknown environments they make use of only a few characteristics and a very limited number of categories. From their research it appears that people only use two out of three independent dimensions to indicate different categories. But which two dimensions people use appears to differ considerably among test subjects. As a consequence, redundant information may help reach more people, but it does not really help improve the performance of individuals.

In everyday traffic, route guidance signs can be easily obscured or obstructed by other (heavy) vehicles. Therefore, the information on route signs should be presented to a driver at least twice in order to reduce their chance of missing a relevant exit on a motorway. After a decision point, drivers like confirmation that they are going in the right direction and by providing them with this information they can drive in a more relaxed way.

Uniformity and Consistency

Uniformity in behavioural rules and applying these rules consistently are both important components to:

- realize uniform road user behaviour;
- meet driver expectations with respect to both road layout and behaviour of other road users;
- reduce human error in everyday road traffic; and
- support the principle of resilience to failure.

A limited set of road delineation situations, combined with unambiguous intended behaviour, can be learned and understood. Uniformity is also important to learn over time, recognize and make the proper use of it, and to develop schemata and prototypical representations to increase information processing efficiency and reduce human error (see also Chapters 2 and 4).

An important example is the solid line that, as a basic rule, drivers are not allowed to cross. However, the current layout of rush-hour lanes on Dutch motorways (see also Figure 3.4) allows road users at certain hours of the day to cross the solid edge lane between the right (slow) lane and the emergency lane (or crossing may even be mandatory because of the general rule of keeping right on a multi-lane road), but this violates the principle of uniformity. This may also devalue the solid line in situations where not crossing is essential for safe use of the road. It may give road users the idea that it does not matter that much and that crossing a solid line is not that serious. If the meaning is not uniform, people may 'value' a rule or a provision in their own 'advantageous' utility – 'incorrect but rewarding violation' (Reason et al. 1998; see also Chapter 9). The opposite situation may occur as well. For example, if one specific type of behaviour is requested – such as stopping – this message is communicated differently at signalized intersections, railway level crossings and movable bridges. One way of communicating that a road user has to stop – for example, a solid red signal preceded by a yellow phase of appropriate duration dependent on approach speed to indicate the transition between driving and stopping – is to be preferred (van der Horst 1988).

Despite the fact that in the Netherlands dynamic speed signs without a red boundary circle on gantries at motorways have the legal status of a speed limit, the majority of road users do not consider them as such (see Figure 3.6). Many drivers think (or want to believe) the signs are indicating an advisory speed.

A speed limit sign with a red boundary circle is unambiguous, much better understood and uniform and consistent in appearance with static speed limit signs

Figure 3.6 Example of dynamic speed signs without a red boundary circle that have the legal status of speed limits in the Netherlands

Figure 3.7 Dynamic and static signs should be uniform in appearance

Figure 3.8 Example of dynamic speed limit sign (80) and Automatic Incident Detection speed sign (70)

– so less prone to (intentional) misinterpretation (see Figure 3.7). In this case the static speed sign is also a kind of fall-back option in case the dynamic signs are not functioning properly.

For this reason, at road section control with a 100 per cent enforcement regime the Variable Message Sign with a red boundary circle is used, whereas for the Automatic Incident Detection system the speed sign without a red circle is used. Figure 3.8 gives an example where both signs are displayed on a motorway gantry.

Forgiving Roadside Environment

Single-vehicle crashes appear to be a significant component of annual road casualties. According to Collin (2000), 33.8 per cent of all fatalities in the European Union in 1998 were the result of single-vehicle collisions with drivers failing to keep their vehicle on the road. The objective of good road design is to provide a roadway that enables drivers to keep their vehicle on the road and, in case a vehicle inadvertently crosses the edge (line), to allow them to recover safely. The reasons for inadvertent roadside encroachments are various and may include drivers trying to avoid a vehicle, object or animal in their travel lane; inattentive driving due to distraction, fatigue, sleep, alcohol or drugs; unexpected effects of weather on pavement conditions; or travelling too fast on a curve. However, roadway design factors – such as traffic lanes that are too narrow given the road function and design speed; substandard curves; or inconsistent road design with unexpected transitions that surprise the driver – may also contribute to the occurrence of run-off-the-road accidents.

Human error cannot be completely avoided and, therefore, geometric road design should enable drivers to timely correct erratic manoeuvres. One measure to induce correcting behaviour may be so-called 'rumble strips' or profiled road markings to alert drivers that they are deviating from their lane. Van der Horst, Vos and Folles (1997) indeed found that profiled road markings on motorways reduced the number of lane-crossings considerably. Räsänen (2005) found that a rumble strip barrier line in a curved section of road was effective in a safety-critical situation where the driver met oncoming traffic. SWOV (2010) refers to Canadian and USA research that shows a 20–60 per cent reduction in run-off-road crashes on sections with rumble strips implemented.

Recent guidelines on roadside design (Thomson et al. 2006) not only refer to obstacle-free zones but also suggest recommendations to improve the condition of hard shoulders such that drivers are able to recover safely from inadvertent lane departure without inviting them to drive on the shoulder on purpose – for example by providing discomfort without losing vehicle control (see Figure 3.9).

An example of behavioural adaptation effects (see Chapter 8) was the positive safety effect of removing reflector posts along Finnish roads (Kallberg 1993) as reflector posts gave drivers more visual guidance at night than needed, with an increase in speed as a consequence.

Figure 3.9 Example of forgiving safe hard shoulder design (CROW, 2004)

Summary and Conclusions

The issue addressed in this chapter is that integrated road systems design should anticipate the potential occurrence of inadvertent aberrations (for example, broken traffic signals, drivers taking the wrong exit, missing a traffic sign) and build in safety measures to ensure that even a breakdown of particular components of the system does not immediately result in accidents (the principle of graceful degradation). The road system design should be robust and meet the sustainable safety principle of forgivingness of the environment and of road users. As human error is a major contributing factor in crashes, the taxonomy of dangerous actions by Reason (1990) and Reason et al. (1998) is discussed in some detail. Several examples are given of how in practice road systems can be made more resilient and robust against human error.

References

Alexander, G.J. and Lunenfeld, H. 1986. *Driver Expectancy in Highway Design and Traffic Operations* (Report FHWA-TO-86-1). Washington, DC: US Department of Transportation, Federal Highway Administration.

Collin, C. 2000. *Statistics in Focus – Transport* (catalogue number CA-NZ-00-003-EN-I).

CROW. 2004. *Handboek veilige inrichting van bermen: Niet-autosnelwegen buiten de bebouwde kom.* (CROW-publicatie 202). Ede: CROW.

EuroStat. Available at: http://epp.eurostat.ec.europa.eu/portal/page/portal/eurostat/home

Elvik, R. 2003. Effects on road safety of converting intersections to roundabouts. In: Transportation Research Record 1847. Washington, DC: Transportation Research Board, 1–10.

Fortuijn, L.G.H. 2009. Turbo roundabouts: Design principles and safety performance. In: *Transportation Research Record 2096 Traffic Safety: Roundabouts, Trucks, Older Drivers, and Traffic Law Enforcement 2009.* Washington, DC: Transportation Research Board, 16–24.

Fortuijn, L.G.H. and Carton, P.J. 2008. *Turbo Circuits: A Well-Tried Concept in a New Guise.* The Hague: Province of Zuid-Holland. Available at: http://www.zuid-holland.nl

Hale, A.R. and Heijer, T. 2006. Defining resilience. In: E. Hollnagel, D.D. Woods and N. Leveson (eds), *Resilience Engineering: Concepts and Precepts.* Aldershot: Ashgate, 35–40.

Horst, A.R.A. van der. 1988. Driver decision making at traffic signals. *Transportation Research Record,* 1172: 93–7.

Horst, A.R.A. van der, Martens, M.H. and Kik, J.J. 2008. *Evaluation of Signalised High-Volume Multi-Lane Turbo Roundabouts: Road Scene Analysis, Road User Survey, and Video-Based Analysis of Road User Behaviour*. Proceedings of the 21st International Co-operation on Theories and Concepts in Traffic Safety (ICTCT) Workshop, Riga, 30–31 October 2008.

Horst, A.R.A. van der, Vos, A. de and Folles, E. 1997. *Lane-Keeping Behaviour at Profiled Road Markings on Motorways: A Before-and-After Study*. Proceedings of the ICTCT 79 Conference, 5 Safety Statistics and Evaluation Studies, 5–7 November 1997. Lund: Lund University of Technology.

Kallberg, V.P. 1993. Reflector posts – signs of danger? *Transportation Research Record,* 1403: 57–66.

Kaptein, N.A., Hattem, S.T. van and Horst, A.R.A. van der. 1998. *Categorization of Road Environments and Driving Speed* (MASTER Deliverable D9). Espoo, Finland: VTT Communities and Infrastructure.

Räsänen, M. 2005. Effects of a rumble strip barrier line on lane keeping in a curve. *Accident Analysis and Prevention,* 37: 575–81.

Rasmussen, J. 1985. Trends in human reliability analysis. *Ergonomics,* 28(8): 1185–95.

Reason, J. 1990. *Human Error*. Cambridge: Cambridge University Press.

Reason, J., Parker, D. and Lawton, R. 1998. Organizational controls and safety: The varieties of rule-related behaviour. *Journal of Occupational and Organizational Psychology,* 71: 289–304.

Reason, J., Manstead, A., Stradling, S., Baxter, J. and Campbell, K. 1990. Errors and violations on the roads: A real distinction? *Ergonomics*, 33(10–11): 1315–32.

Sagberg, F. 2007. *A Methodological Study of the Driving Reliability and Error Analysis Method (DREAM)* (TØI report 912/2007). Oslo: TØI Institute for Transport Economics.

Salmon, P.M., Regan, M.A. and Johnston, I. 2006. *Human Error and Road Transport: Phase Two – A Framework for an Error Tolerant Road Transport System* (Report No. 257). Clayton, Victoria: Monash University Accident Research Centre.

SWOV. 2010. Factsheet *Safe Road Shoulders*. Leidschendam: SWOV Institute for Road Safety Research.

Thomson, R., Fagerlind, H., Martinez, A.V., Amenguel, A., Naing, C., Hill, J., Hoschopf, H., Dupre, G., Bisson, O., Kelkka, M., Horst, A.R.A. van der and Garcia, J. 2006. *European Best Practice for Roadside Design: Guidelines for Roadside Infrastructure on New and Existing Roads* (RISER Deliverable D6). Gothenburg: Chalmers University of Technology.

Vonk, T., Rooijen, T. van, Hogema, J. and Feenstra, P. 2007. *Do Navigation Systems Improve Traffic Safety?* (TNO Report 2007-D-R0048/B). Delft: TNO Mobility and Logistics.

Wegman, F. and Aarts, L. (eds). 2006. *Advancing Sustainable Safety: National Road Safety Outlook for 2005–2020*. Leidschendam: SWOV Institute for Road Safety Research.

Chapter 4
The Performance of Road Users: Hierarchical Task Levels

Richard van der Horst

Driving involves three hierarchical task levels: viz. navigation, guidance and control. For each level, task load varies depending on time and place. In road system design it is important to distinguish between all three levels.

Theoretical Background

Several studies on accident causation factors indicate that human-related errors are to be considered as a major contributing factor (Mansfield et al. 2007). Rumar (1985) gives a still highly illustrative example of the relative contribution of the three main components of the road traffic system, separately and in combination (see Figure 4.1).

As about one-quarter of crashes originate from a combination of road and road user factors, the interaction between road environment and road user is an important area to consider for ensuring the safe and operationally efficient mobility of all road users. However, road user factors such as inattention/distraction or inappropriate speed choice – which, according to Mansfield et al. (2007), are attributed to 82 per cent of this group – may also relate directly or indirectly to the road environment. Driving a motor vehicle is for many people the most complex and potentially dangerous task they will perform during their lives (Regan, Lee and Young 2008) and consists of the complex interaction between the road and traffic environment, the vehicle and the road user. From the perspective of the

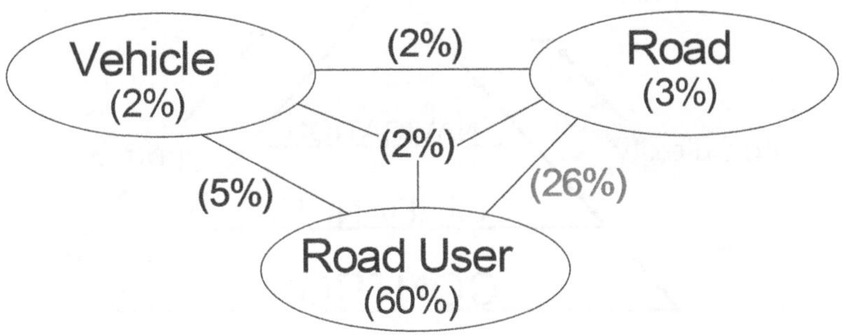

Figure 4.1 Contributing factors in road crashes (from Rumar 1985)

road user, roadway design (including traffic management) therefore considerably influences the tasks drivers have to perform. For this reason it does make sense to consider road and traffic management with the driving task in mind. After a description of the driving task and of modelling driver behaviour, this chapter will deal with the consequences with respect to effects on driver behaviour and the implications for roadway systems design.

The Driving Task

In the literature, the task analysis for driving a car is well documented. A frequently used conceptual model of the driving task consists of three hierarchically ordered levels: *navigation*, *guidance* and *control* (Allen, Lunenfeld and Alexander 1971). Tasks at the navigation level refer to activities related to planning and executing a trip from origin to destination. The need for processing information only occurs occasionally, with intervals ranging from a few minutes to hours. Decisions made at this level are giving input to the next hierarchical level. The guidance level refers to tasks dealing with the interaction with both the environment (roadway, traffic signs/signals etc.) and other road users (overtaking, being overtaken, yielding, stopping etc.). Activity is required rather frequently, with intervals of a few seconds to a few minutes. This task level refers to task elements such as speed choice and lane choice and provides the input to the lowest task level. At the control level the motion of the vehicle is controlled in longitudinal and lateral direction, and information has to be processed frequently, ranging from intermittent activities every few seconds to almost continuous control. Alexander and Lunenfeld (1986) visualized the relationship between the levels as a set of nested triangles (Figure 4.2), hierarchically ordered from a low to a high level with an increasing *complexity* and from high to low with an increasing urgency (*primacy*). For example, a flat tyre or a sudden heavy gust of wind will immediately interrupt activities at the navigation level and focus all attention on the control level, since getting lost has less severe consequences than running off the road.

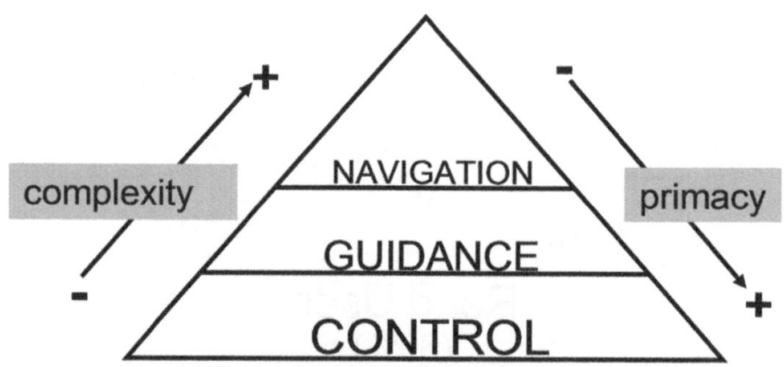

Figure 4.2 The three hierarchical levels of the driving task (according to Alexander and Lunenfeld 1986)

At each level of the driving task, the successive steps of the cycle of human information processing – for example, perception, processing, decision making and action – take place. All these processes take time and have different time constants. Moreover, the way a driver performs these tasks and the time it takes strongly depend on the routine they have developed in task performance. Rasmussen (1985) distinguishes three levels of task performance: *knowledge-based*, *rule-based* and *skill-based.* The highest level (knowledge-based) refers frequently to new situations (for example, finding the best route to a new destination) or situations that occur frequently in themselves but of which the driver still has little experience. The choice of behaviour depends on interpretation and deductive reasoning. If a driver is frequently confronted with a given situation, after some time a rule develops about how to deal with that situation, and recognising that situation leads to appropriate behaviour in a 'programmed' manner without the 'need' to understand exactly what is going on. Once a driver has recognised the situation and a 'programme' has been selected, task execution occurs somewhat automatically.

Skill-based tasks are conducted automatically and very quickly: incoming information automatically results in the right behaviour without any cognitive control. Theeuwes (1993) introduced a good three-dimensional representation of the driving task (as shown in Figure 4.3), with the three hierarchical levels, the stages in human information processing and the different levels of task performance combined. Usually, for an experienced driver the relationship between task hierarchical levels and task performance levels is diagonal: viz. tasks at the strategic level are conducted at a knowledge-based level; tasks on the manoeuvring level are linked to rule-based task performance and control level tasks to skill-based task performance. But this is not necessarily a one-to-one relationship. For each individual driver, the way the task at each of the three hierarchical levels of

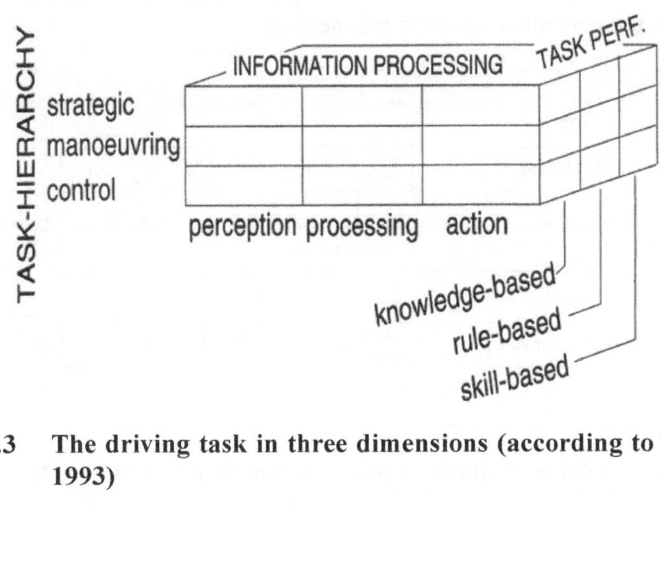

Figure 4.3 The driving task in three dimensions (according to Theeuwes 1993)

the driving task is conducted may shift over time. For example, a novice driver initially will conduct the control task of the vehicle (shifting gears, accelerating, braking, steering etc.) at a knowledge-based task performance level that requires most of the attentional resources (see below). However, as the skill of controlling the vehicle develops by training and practice, task performance will shift to a more automatic level (skill-based), with no or very little cognitive attention needed any more. Similar processes may take place at the other levels of the driving task.

Modelling Driver Behaviour

As indicated above, to a large extent driving behaviour is determined by various external feedback factors from the environment (roadway characteristics, behaviour of other road users, weather and visibility conditions) as perceived by the driver. But internal factors also play a role in the way drivers perform their task. A frequently used theoretical model is the Theory of Planned Behaviour proposed by Ajzen (1985). This model indicates that the *motivation* of drivers resulting in certain behaviour is based upon their *intentions* that in turn are determined by *attitude*, *subjective norm* and *perceived behavioural control*. The human information processing elements will all be influenced by drivers' intentions. An overall model that covers these aspects is given in Figure 4.4 (van der Horst 1998).

The human information processing model of Wickens (1984) illustrates that humans have a limited information processing capacity (see Figure 4.5). This limitation will become apparent when a task involves a high attentional demand. For example, if driving involves a task at a knowledge-based level (such as navigating in an unknown city without an electronic navigation system) it is expected that most resources will be devoted to this task level. It is likely that lower-level tasks will suffer. For example, a driver involved in a highly complex knowledge-based task may start swerving (which is a task conducted at the lowest level). The schemata as mentioned in Chapter 2 also fit well in this model and reflect how the content is stored in the memory.

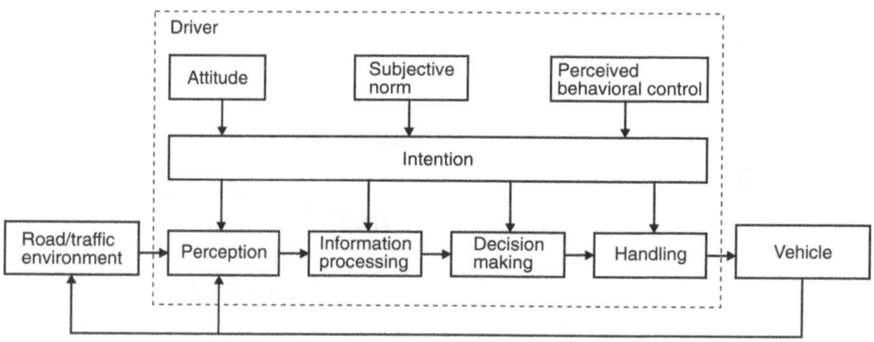

Figure 4.4 A driver behaviour model (according to van der Horst 1998)

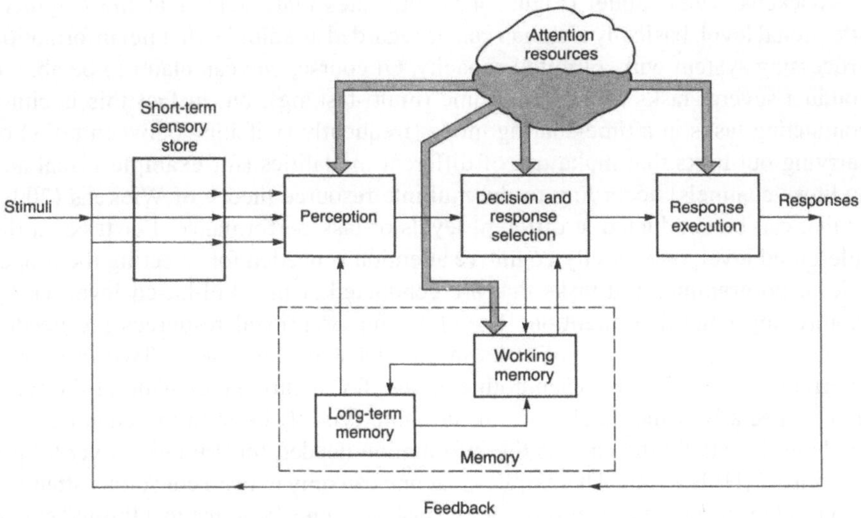

Figure 4.5 Human information processing model of Wickens (1984)

Wickens' attention resources reflect a potential bottle-neck (limited capacity) in the information processing cycle due to limited attention resources. Even though over the years several authors have located this bottle-neck at different locations in their models, they all agree on the presence of a bottle-neck somewhere in this cycle (Martens 2007). See also Chapter 5 on workload.

Effects on Driver Behaviour

The way a driver performs the driving task also strongly depends on their driving experience. As mentioned, a novice just taking their first driving lessons conducts the control level of the driving task (vehicle handling) mainly in the knowledge-based performance mode that takes most of the available cognitive attention resources, with the consequence that little attention can be paid to the other subtasks. As the driver gains experience of handling the car, task performance shifts more to a skill-based performance mode. Similar processes at the navigation level may occur when repeatedly driving the same route. On the one hand, from the point of view of task efficiency and effectiveness task execution at a more or less automated level is highly advantageous. But on the other hand, in situations for which changes in handling are needed (for example driving a vehicle with an automatic transmission while used to a hand-shift gear with a clutch pedal) it may result in severe errors. Processes that are conducted at the skill-based level are difficult to interrupt or suppress, and this takes extra time.

Wickens' 1984 model (Figure 4.5) illustrates that, at least at the cognitive attentional level, basically a human can be regarded as a single-channel information processing system with a limited capacity. Of course, one can claim to be able to conduct several tasks at the same time (multi-tasking), but in fact this is either conducting tasks in a time-sharing mode (frequently switching between tasks) or carrying out tasks that make use of different modalities (for example visual and auditory channels) according to the multiple resource theory of Wickens (2002) or that can be conducted at different levels of task performance. For tasks at the rule-based level, only briefly cognitive attention is needed for selecting the proper rule or programme; and tasks that are conducted at the skill-based level hardly require any cognitive attention. Therefore, no attentional resources are needed for conducting these types of tasks in parallel with other tasks. Two tasks will interfere with each other when both compete for attentional demand at the same time, especially if they make use of the same (visual) information channel. And the latter is usually the case as the information needed for driving is over 90 per cent visual (Hills 1980). This implies that one can only expend conscious attention on visual information processing for one task at a time, be it that to a limited extent the concept of short-term sensory store or temporary storage in working memory allows for processing stimuli that are presented at the same time, or one after the other. The consequence of the single-channel view is that one can only give cognitive attention to hierarchical driving task levels one at a time and that there is the risk of overload, with severe task performance consequences, if the task to be conducted requires conscious attention – for example at the manoeuvring level to decide whether or not to overtake in a complex traffic situation or dealing with merging traffic at an oncoming exit etc.

Information needed for conducting tasks at one driving task level should not compete with information needed for conducting tasks at another driving task level at the same time if conscious attention is required for both tasks. The consequence for roadway design and traffic management systems is that one needs to separate in place and time information requirements that are needed to execute the different task levels. For example, at a complex motorway interchange drivers first need information on directional choices at the guidance level in order to decide whether or not they should take action. Then they have to orientate themselves to prepare for the desired lane choice at the guidance level in connection with the outcome of the decision at the navigation level. Given a lane choice, the effectuation of this choice has to be dealt with in interaction with the presence of other traffic, together with the choice of an appropriate speed; and finally this speed and lane manoeuvre has to be conducted at the control level. Each of these steps takes time and should be made possible by providing the information in the right order and with enough time to accommodate them. The limited human information processing capacity at the knowledge-based level also implies that distraction (for example paying attention to non-driving related information or tasks) can result in dangerous situations as one misses relevant information, does not know which direction or lane to take or overlooks other traffic.

Situation awareness (including perception of relevant elements in the current situation, comprehension of the current situation and projection of the future situation); *workload* (too much is bad and results in stress, too little is bad because it results in a low arousal level of the driver (see Chapter 5); and *attention* (the stage preceding the actual processing by the senses) appear to be crucial human factors aspects in road safety. As an example of the latter factor, in the USA a large-scale study was completed in which 100 car drivers were observed in great detail for a one-year period in their vehicles while driving to get more insight into the pre-crash phase of collisions – the so-called '100-car naturalistic study' (Neale et al. 2005). This study registered 82 crashes and 761 near-crashes (defined as a conflict situation requiring a rapid, severe evasive manoeuvre to avoid a crash). Figure 4.6 gives an example of a categorisation of events in which the driver was attentive or inattentive. Driver inattention to the forward roadway was found to be the primary contributing factor in most crashes, with 93 per cent of the rear-end crashes involving driver inattention. Driver inattention included such things as eating, writing, conversing with passengers, secondary task engagement, fatigue; driving-related inattention to the forward roadway by looking in the rear-view mirror; and non-specific eye-glances away from the forward roadway to objects inside or outside the vehicle.

These categories were extracted manually by data reductionists from video footage. It appeared that a majority of events (78 per cent of the crashes and 65 per cent of the near-crashes) had driver inattention as a contributing factor. The sources of inattention that generally contributed to the highest percentages of events were wireless devices, internal distractions and passenger-related secondary tasks.

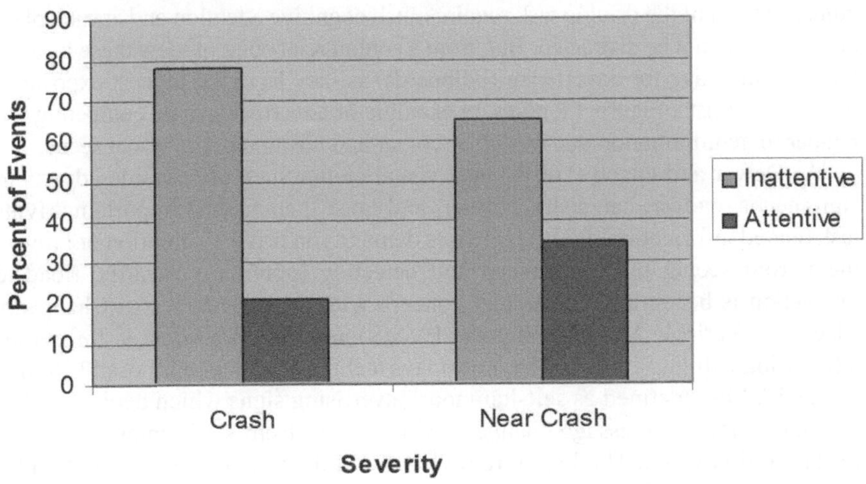

Figure 4.6 Percentage of crashes and near-crashes with driver inattention/ distraction as a contributing factor (Neale et al. 2005)

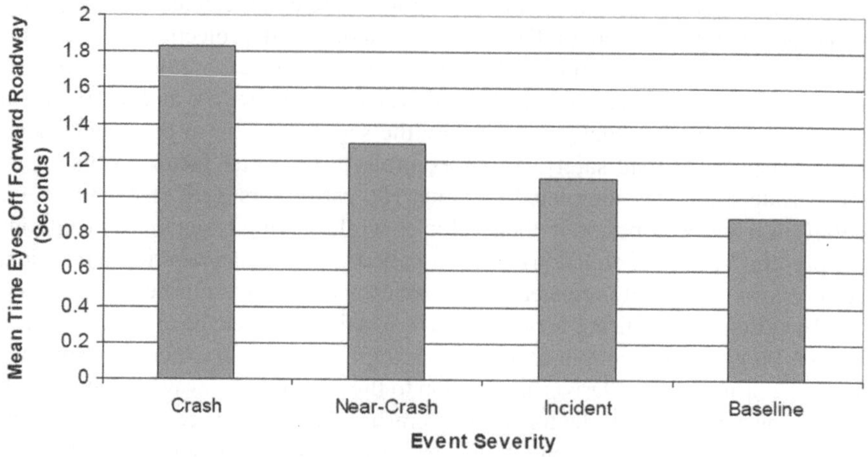

Figure 4.7 Time eyes off forward roadway as a function of event severity (Klauer et al. 2006)

In the same study it was concluded that in crash events the duration of time the drivers' eyes were off the road was significantly larger than in the baseline events, providing a design rule that glances for in-car tasks should not take longer than 1.8 seconds in any case (see Figure 4.7).

Together with the time dependency of required information at different task levels of the driving task, it is clear that distraction should be avoided at important decision points in the road network for all three levels of the driving task as much as possible. Especially, at major motorway interchanges both the navigation and the guidance level of the driving task require much cognitive attention and at such places drivers should not be distracted. But, from a commercial point of view these locations are very attractive for advertising (billboards) as they have the highest exposure to passing motorists. Figure 4.8 gives an example of advertising signs competing with regular route information due to similar colour and position in the visual scene.

As Dewar and Olson (2001) state, visual clutter such as roadside advertising signs near a roadway can distract drivers and cause them to miss important driving-task related information. It also increases demands on drivers' attention or prevents them from seeing the road ahead and detecting (potential) hazards. Roadside distraction is becoming a particular concern with the appearance of Changeable Electronic Variable Message Signs (CEVMS) and digital billboards for outdoor advertising (Molino et al. 2009). Nowadays technologies make it possible to have huge CEVMS – defined as self-luminous advertising signs which depict any kind of light, colour or message change – which range from static image sequences to full motion video. The literature review by Molino et al. (2009) indicated that the current body of knowledge represents an inconclusive scientific result with regard to demonstrating detrimental driver safety effects due to CEVMS exposure. They proposed a three-stage research programme, including determination of

Figure 4.8 Example of advertising signs along a Dutch motorway

distraction, basis for possible regulation and relationship of roadside distraction to crashes. For the moment, the Netherlands has a rather restricted policy for application of CEVMS along Dutch motorways.

Implications for Roadway Systems Design

The distinction between the different levels of the driving task and the notion that each level has its own human information processing requirements in the knowledge-based performance mode by separating both information provision and task execution in place and time helps make road design and traffic management systems more manageable for the driver. Also the information processing time increases if driver expectation is not met: see, for example, the difference in detection distance if a target destination one is looking for is or is not on a route sign (see Chapter 6, Figure 6.4). Road measures intended for use by one of the task levels should fit the usual level of task performance associated with it – such as traffic signs indicating that one should follow a different colour of lane markings – may not be sufficient to actually change lane keeping behaviour that is conducted at a skill-based level (see Figure 4.9).

Another example is that drivers experience tunnel entrances as extra loading situations that may result in stress and uncertainty (Verwey 1995). Visual scan patterns and physiological measures indicate that drivers start preparing for a

Figure 4.9 Regular white lane markings still seem dominant for lane keeping over temporary yellow lane markings in work zones

tunnel entrance 200 metres in advance. At such locations where the task at the guidance level requires extra attention one should not display information that is relevant for the navigation level of the driving task. An example of a situation that should be avoided is given in Figure 4.10, where the route sign for the exit directly after the tunnel can easily be missed, as one pays attention to negotiating the tunnel mouth itself.

Figure 4.10 Example of a less desirable combination of information at both the navigation and guidance level of the driving task

Figure 4.11 Bad example of a combination of information given to drivers at the same location

An example of an unnecessary and poor combination of different information sources is given in Figure 4.11. The green panel with non-driving-task relevant general maintenance information should not compete with information on dynamic route information panels (DRIP), whereas the distance between DRIP,

Figure 4.12 **Example of the Dutch motorway route signing system with a pre-warning at 1,200 metres (top), a repetition sign at 600 metres (middle) and confirmation at the decision point for both through-going and exiting traffic (bottom)**

route information and decision point should also be much greater than 300 metres. Too much visual information in the same place should be avoided in any case to prevent visual overload and competition among the different levels of the driving task.

Figure 4.12 shows a good example of the adequate and timely separation of information as the standard route signing on Dutch motorways, which features:

- a pre-warning signal at 1,200 metres;
- a repetition sign at 600 metres (drivers may have missed the sign along the road due to obstruction by a preceding truck, did not fully catch the information on the first sign or were momentarily occupied by a lane change manoeuvre by another road user, etc.);
- a confirmation sign at the exit itself (drivers like confirmation that they have made the right choice).

In this manner drivers have time to decide whether or not they have to exit before subsequently starting to perform the task at the guidance level to make a proper lane and path choice.

Since 2007, the approach for route signing in the Netherlands has been changed, with lane indicator arrows on route signs pointing up instead of down, more or less following German guidelines. At interchanges with more than one exit lane, this type of lane indicator on route signs helps promote better use of all lanes as it communicates to road users that they can also use the right through lane for a while to exit (see Figure 4.13 on the following page). So, once the decision has been made to turn right, the arrows give additional information on how to act at the guidance level of the driving task.

Summary and Conclusions

This chapter provides an overview of the main theories and models of driver behaviour and performance. The interaction between the road environment and the road user is an important area to consider in order to ensure a safe and operationally efficient mobility of all road users. Car driving is a complex task that consists of three hierarchically ordered task levels: navigation, guidance and control. Each level requires its own information input and the successive stages in information processing of perception, processing, decision making and action. As a human basically can be regarded as a single-channel information processing system with a limited capacity, it is important to separate the information provided to a driver for each level in place and time as much as possible. Situational awareness, workload and attention are major human factors aspects to take into account. Several examples of implications for road systems design illustrate the applicability of this approach.

Figure 4.13 Example of new route signage with additional information on the route panels for use at the guidance level of the driving task

References

Ajzen, I. 1985. From intentions to actions: A theory of planned behaviour. In: J. Kuhl and B. Beckmann (eds), *Action Control: From Cognition to Behaviour*. Berlin: Springer-Verlag, 11–39.

Alexander, G.J. and Lunenfeld, H. 1986. *Driver Expectancy in Highway Design and Traffic Operations* (Report FHWA-TO-86-1). Washington, DC: US Department of Transportation, Federal Highway Administration.

Allen, T.M., Lunenfeld, H. and Alexander, G.J. 1971. Driver information needs. *Highway Research Record*, 366: 102–15.

Dewar, R.E. and Olson, P.L. 2001. *Human Factors in Traffic Safety*. Tucson, AZ: Lawyers and Judges Publishing Company, Inc.

Hills, B.L. 1980. Vision, visibility and perception in driving. *Perception*, 9: 183–216.

Horst, A.R.A. van der. 1998. *Factors Influencing Drivers' Speed Behaviour and Adaptation* (TNO Report TM-98-D006). Soesterberg: TNO Human Factors Research Institute.

Horst, A.R.A. van der and Hagenzieker, M.P. (eds). 2002. *Report of the European Workshop on International Human Factors Guidelines for Road Systems* (TNO Report TM-02-D009). Soesterberg: TNO Human Factors Research Institute.

Klauer, S.G., Dingus, T.A., Neale, V.L., Sudweeks, J.D. and Ramsey, D.J. 2006. *The Impact of Driver Inattention On Near-Crash/Crash Risk.* (NHTSA report DOT HS 810 594). Washington, DC: National Highway Traffic Safety Administration.

Kahneman, D. 1973. *Attention and Effort*. Englewood Cliffs, NJ: Prentice Hall.

Mansfield, H., Bunting, A., Martens, M. and Horst, A.R.A. van der. 2007. *Analysis of the On the Spot (OTS) Road Accident Database* (Road Research Report No. 80). London: Department for Transport.

Martens, M.H. 2007. *The Failure to Act upon Important Information: Where Do Things Go Wrong?* PhD thesis. Amsterdam: Vrije Universiteit.

Molino, J.A., Wachtel, J., Farbry, J.E., Hermosillo, M.B. and Granda, T.M. 2009. *The Effects of Commercial Electronic Variable Message Signs (CEVMS) on Driver Attention and Distraction: An Update* (Publication No. FHWA-HRT-09-018). Washington, DC: US Department of Transportation, Federal Highway Administration.

Neale, V.L., Dingus, T.A., Klauer, S.G., Sudweeks, J. and Goodman, M. 2005. *An Overview of the 100-Car Naturalistic Study and Findings*. Proceedings of the 19th Enhanced Safety Vehicles Conference, Washington, DC, June 2005 (Paper No. 05-0400).

Rasmussen, J. 1985. Trends in human reliability analysis. *Ergonomics*, 28(8): 1185–95.

Reason, J. 1990. *Human Error*. Cambridge: Cambridge University Press.

Regan, M.A., Lee, J.D. and Young, K.L. (eds). 2008. *Driver Distraction: Theories, Effects, and Mitigation*. Boca Raton, FL: CRC Press/Taylor and Francis Group.

Rumar, K. 1985. The role of perceptual and cognitive filters in observed behaviour. In: L. Evans and R.C. Schwing (eds), *Human Behaviour and Traffic Safety*. New York: Plenum Press, 151–70.

Theeuwes, J. 1993. *Visual Attention and Driving Behaviour*. Proceedings of the International Seminar Human Factors in Road Traffic, 5–6 April 1993, Braga, Portugal: Universidade do Minho, 103–23.

Verwey, W.B. 1995. *Effects of Tunnel Entrances on Driver's Physiological Condition and Performance: An Exploratory Study* (TNO Report TNO-TM 1995 C-19). Soesterberg: TNO Human Factors Research Institute.

Wickens, C.D. 2002. Multiple resources and performance prediction. *Theoretical Issues in Ergonomics Science*, 3(2): 150–77.

Wickens, C.D. 1984. *Engineering Psychology and Human Performance*. Columbus, OH: Bell and Howell.

Chapter 5
Workload Management

Jan Theeuwes

When approaching task overload, road users will adapt their behaviour such that the load becomes lighter. Additional tasks will be ignored or skipped. When actual task overload occurs, rationale behaviour may cease to exist and drivers may fall back to basic, rather reflexive behaviour tendencies.

Theoretical Background

The concept of task overload is related to mental workload, which refers to the information processing demands imposed by the tasks that need to be executed (for example, De Waard 1996; see Figure 5.2). Task overload 'is invoked to account for those aspects of the interaction between a person and a task that cause task demands to exceed the person's capacity to deliver' (Gopher and Donchin 1986: 3). As is clear from this definition, mental workload and the extent to which this may result in task overload depends on the task requirements and the capacity of the person. Different people have different capacities, so the extent to which someone experiences overload depends very much on the characteristics of that person. It is also important to realize that different tasks or combinations of tasks may call upon different capacities. For example, tasks that are highly automatized may call upon few, if any, processing demands.

The underlying idea is that people have a limited amount of processing resources available. If the task demands exceed the amount of processing resources available one speaks of 'task overload'. Resources can be considered as a set of commodities available for the human information processing system. Some theories have suggested that there is one undivided resource available (Kahneman 1973); others have suggested that there are different resources for different task components. For example, the multiple resources theory of Wickens (1984) assumes that there are resources available for the different modalities (visual versus auditory), processing codes (spatial versus verbal) and responses (manual versus vocal). Wickens (1984) referred to workload with the phrase 'so much to do, so little time', stressing that often time pressure is one of the most evident features that people experience when they are overloaded.

Different Components of Workload

Workload is a complex phenomenon that has various effects on physiology, performance and subjective feelings. Figure 5.1 gives an overview of the effects

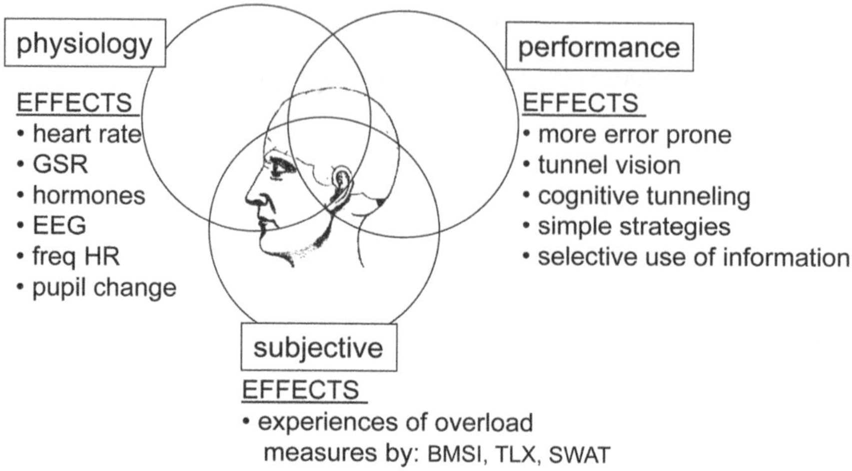

Figure 5.1 The various components of workload (physiology, performance and subjective experience) and their effects

of these different components of workload. As is clear from the figure, when people experience workload there are several physiological reactions of the body, including:

- an increase in heart rate;
- a reduction of heart rate variability;
- an increase in the size of the pupil;
- increased sweat emission (galvanic skin response);
- release of stress hormones (for example, cortisol);
- changes in event-related brain potential (ERP).

At the same time, people experience subjectively higher demands which may be related to various aspects of the task, such as time pressure, frustration level, effort and the mental activity needed. Well-known subjective measures of workload are, for example, the Task Load Index (TLX), Subjective Workload Assessment Techniques (SWAT) or the BMSI (a Dutch version of a subjective workload assessment technique). One task that is often used to measure workload in experimental settings is the peripheral detection task (PDT) (see, for example, Van Winsum, Martens and Herland, 1999). This task consists of the driver detecting a small stimulus (for example, a red square) presented on the windscreen of the car, usually in the periphery at 10–20 degrees. The driver has to respond as quickly as possible (for example by pressing a microswitch) as soon as they detect the stimulus. The idea is that reaction time (RT) goes up when the driver is occupied with driving. The greater the RT increase the larger the momentary workload. Typically, in conditions of overload, task performance is affected such that people make more errors, use simple strategies and reduce their visual field – sometimes referred to as tunnel vision.

Demand–Performance Relationship

To understand overload one has to consider the demand–performance relationship (see Figure 5.2). As long the demands of the task remain within the attentional capacity of the person performing the task, no overload is experienced. If the task demands are not high, the operator may experience normal workload. However, if the workload keeps increasing, performance may suffer unless people invest more effort in keeping the task at a high level (see Figure 5.2).

Performance begins to decline as the task demands approach the maximum resources available – that is, with maximum effort. At this point task performance deteriorates even though, with maximum effort, the operator tries to keep the task at a high level. This is when one speaks of task overload. It is important to realize that the task demands can also be too low. As is clear from Figure 5.2, it is possible that when the task demands are very low it is hard to keep performance at an adequate level. In this case one speaks of 'underload'. Note that when the task demands are very low, effort invested by the operator cannot improve performance. It has been suggested that low task demands reduce the arousal level of the operator to such a low level that the attentional capacity to perform is reduced (for example, Young and Stanton 1997). Underload is a condition that may occur in situations where a

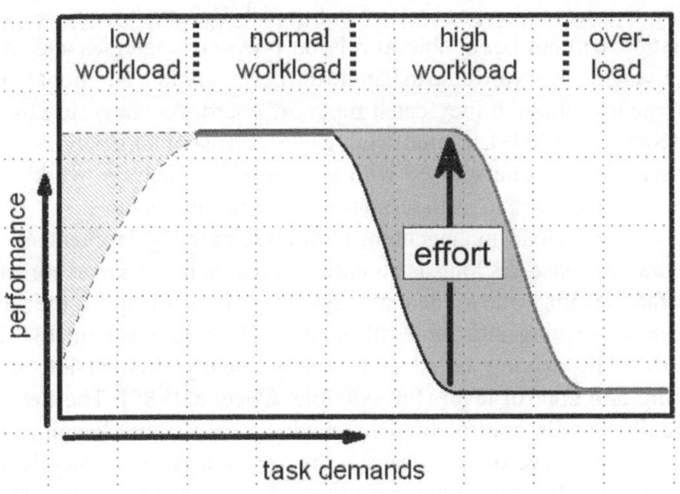

Figure 5.2 The task demand performance relationship

Note: When the task demands are too low there is underload; when there is high workload, by investing more effort one can keep performance at an adequate level. When the task demands approach the maximum resources available one speaks of overload.

Source: From Veltman and Jansen 2003.

task is highly automatized. For example, incident detection is reduced when the operator only has to monitor the processes rather than actively being involved in them (for example, Wickens and Kessel 1981).

Workload During Driving

For a novice driver all aspects of the driving task may be demanding. Even low-level control of the vehicle – such a lane keeping, curve negotiation and speed control – may pose difficulties for novice drivers. Providing additional information to these drivers, sometimes even task instructions, may immediately result in task overload and possibly even panic. It is the same process that children go through when learning to ride a bike. The mere control of the bike is already so demanding that any remark from the parent may immediately result in a fall. Because the various subtasks of the driving task become automated, after training what seemed to be uncoordinated subtasks become fully integrated into a coherent driving task. The experienced driver manages the interrelated set of 'overlearned' subtasks, dealing smoothly with low-level vehicle control, planning manoeuvres, primary hazard monitoring as well as with secondary tasks such as radio tuning and talking to passengers. Under normal circumstances, the experienced driver can perform all primary and secondary tasks without problem.

However, when an emergency situation occurs problems may become apparent (for example, Hancock, Lesch and Simmons 2003). When something unexpected happens the overlearned capacities are not useful. Drivers have to respond to the novel situation without being able to rely on previous strategies and overlearned reactions. Even if drivers are able to handle the new situation, additional distractions to the emergency situation may catch them off guard. As suggested by Hancock, Lesch and Simmons (2003), normal driving may consist of long periods of relatively low workload driving interspersed with moments in which the workload rises to extreme heights (referred to as the 'moment of terror and panic').

The key to workload management is to anticipate the moment in which the workload may increase. As long as potential increases in workload are anticipated, copying strategies may allow adequate responses. In this respect, the well-known classification of driving into three different levels is relevant for the discussion on workload. The driving task can be hierarchically divided into a strategic, manoeuvring and control level (for example, Michon 1985). The lowest control level basically entails control of the vehicle, such as lane keeping and speed control. For experienced drivers, tasks at the lowest level are basically automated such that they hardly require any resources. As noted, novice drivers still have to devote a fair amount of resources to simple control tasks, making them much more vulnerable to task overload when the task demands increase. Moreover, increase in workload is mostly associated with tasks that need to be performed at the manoeuvring level involving interactions with other road users. Especially in busy traffic, the necessary interactions with other road users become more demanding. Verwey (2000) indicates that visual rather than cognitive workload increases by tasks executed at the manoeuvring level. Importantly, Verwey (2000) also points out that curve negotiation, which is a control task, remains visually demanding. For example, after exiting a motorway, there are often relatively large and prolonged curves that require high visual capacity.

In order to reduce workload, it appears that predictability of the road environment is crucial. As outlined by Verwey (2000), it seems that driving behaviour in a consistent road environment automates because 'off-the-shelf' memory representations for similar situations develop with practice. As outlined by Theeuwes and Godthelp (2005), consistent, easily recognizable road environments – self-explaining roads (SERs) – allow the development of consistent behaviour repertoires. When a particular situation occurs, the appropriate repertoires can be loaded from long-term memory in a single step and require little attention when executed. This is consistent with the idea that the driving skill is based on the availability of a large set of standard behaviours which can be selected and executed without much effort. The crucial point is that the situation should be predictable and recognizable and that the standard behaviour patterns should be appropriate for the situation. Uncertainty will contribute strongly to workload increase.

Particular factors, such as adverse weather conditions or unclear road markings, can increase workload during a whole driving trip. In addition to these longer and enduring increases in workload, it is important to realize that in driving the workload can suddenly increase dramatically, exceeding the maximum available resources quickly. This occurs when an expected event happens during driving. Because of the enormous time pressure to act immediately, task load may rise very quickly. One can think of situations such as a sudden traffic jam, a roadblock, a blowout, another car drifting into the lane, unexpected road deviations etc. Workload can also increase because of uncertainty about how to behave appropriately. For example, in the Netherlands in the rush hour, when the emergency lane is used as a regular lane, drivers may be unsure whether they should move into that lane, whether they can exit that lane when needed, whether they are allowed to overtake other cars etc. All situations in which standard, easily retrievable behaviour repertoires are not appropriate will result in increased workload. Table 5.1 on the following page provides some examples of factors that may increase the workload experienced.

Task Prioritization and Compensation

One of the most common copying strategies when drivers are faced with task demands that exceed available resources is task prioritization. When demands are too high, drivers try to reduce them by using alternative strategies. The most common strategy is to reduce additional subtasks that are not necessary for the main task. For example, at times of overload drivers stop talking on the phone, stop listening to the radio, stay in their lane and reduce speed. By skipping and ignoring all tasks that are not immediately necessary, resources are freed up such that the main task

Table 5.1 Examples of conditions that are likely to increase driver workload

Road environment	Car environment	Driver	Short workload increase
• Unfamiliar route • Unexpected design of roads • Unclear signing • Unclear lane marking • Unclear what behaviour is required • Visual distraction • Low sunlight • Dusk and dawn • Rain (glare) • Headlamp glare • Strong light changes • Tunnel entrances	• Mobile phone use • Handling radio • Eating/talking • Handling children • In-car email/SMS	• Driving experience • Gender • Age • Alcohol, drugs • Fatigue	• Multiple lane changes • Limited space manoeuvring • Merging • Exiting • Curve negotiating • Sudden traffic jam • Roadblocks • Work zone

performance is protected from decline. This in itself is a good strategy but problems may arise when particular vital aspects of the driving task are ignored to free up resources. For example, drivers may choose not to check mirrors, not to look over their shoulder or scan for potential hazards in the immediate environment. Ignoring vital aspects of the driving task may ultimately result in collisions. As outlined in Figure 5.1, tunnel vision may be one of the results of high workload. Tunnel vision is the tendency to only look straight ahead and not to monitor the visual periphery for hazards. Cognitive tunnelling is a similar strategy in which drivers devote their attention exclusively to one aspect of the driving task that they think is going wrong. For example, drivers who are unsure of where they are and when to exit a roadway start scanning exclusively for route information signs while ignoring other aspects of the driving task. Other strategies chosen when task overload becomes eminent are to selectively use particular information and to use relatively simple strategies such as exclusively directing attention to the car in front when driving in fog or rainy conditions. Even though some strategies may work, dangerous situations may occur when important aspects of the driving task are simply ignored.

Another situation that can be dangerous is when task demands are increasing while the driver does not notice. If drivers are not aware of the increasing demands of the situation they will not compensate accordingly. In other words, there is no need to adapt behaviour (for example by slowing down) because the increasing demands are simply not noticed by the driver. Experienced drivers are better than novices at recognizing that a hazardous situation may develop and therefore start to take action before a very high overload situation occurs.

When drivers are suddenly confronted by high overload they may abandon adequate strategies to address the overload and instead fall back on rather reflexive behaviour tendencies. Drivers who are overloaded suddenly may panic; decelerate severely; take an improper route alternative; make late or erratic manoeuvres (sudden lane changes); ignore critical information; fail to monitor other traffic; or have excessive 'eyes-off-the-road' time. These erratic behaviours have obvious safety consequences.

How to Design Environments That Do Not Induce High Workload

From the analysis above it is clear that workload increase may not be a problem if:

1. it is gradual and expected; and
2. drivers are aware that they are distracted.

If it is gradual and expected and drivers are aware, then they may compensate for it – for example by slowing down or using other adaptive strategies. Even though these strategies may work, it should be realized that any type of distraction may

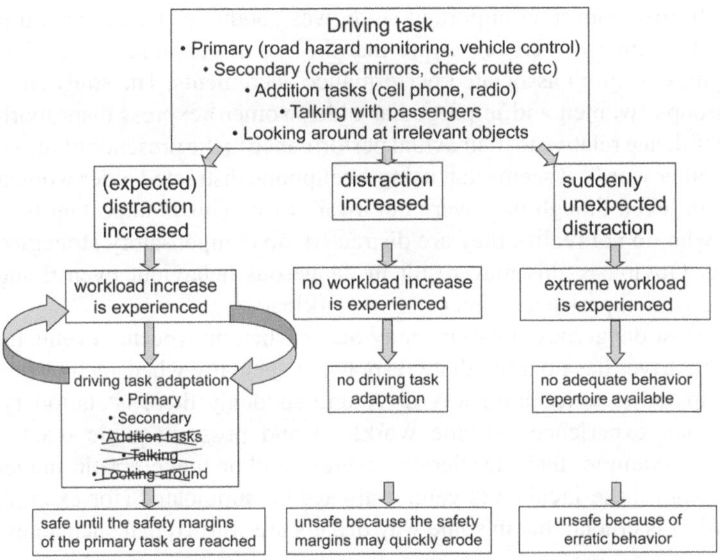

Figure 5.3 Overview of the effects of distraction while driving

Note: Left – If the distraction is anticipated and drivers are aware of the increase in workload they may adapt their behaviour such that the workload stays within acceptable limits. Drivers will choose to first skip those tasks that are not immediately relevant for driving. Only when the workload remains high and adaptation no longer helps will safety margins erode and collisions likely occur. Middle – If the distraction is increased but drivers are not aware of this, they may fail to adapt their behaviour. Safety margins will be reached quickly without drivers realizing this. Right – Sudden, unexpected distraction may result in unsafe and erratic behaviour because no adequate behaviour is readily available.

erode the safety margin in driving compared to the performance of fully responsive and undistracted drivers.

The obvious way to ensure that task workload does not increase is to design road environments that are predictive and easy recognizable. As outlined elsewhere (Theeuwes and Godthelp 1995), one should design roads that are self-explaining. Given a particular road environment, the driver should be able to rely on overlearned standard behavioural repertoires. Deviations from overlearned repertoires are suboptimal. For example, to reduce congestion currently in the Netherlands, depending on the time of the day, drivers may use the emergency lane as regular additional lane. When this rule is in effect, drivers can no longer use the regular behaviour repertoires but should adapt their behaviour in some way or another. Because this occurs only on some roads during some certain of the day, this is suboptimal and will increase workload. Other examples are detours due to roadworks, adverse weather conditions, low sunlight etc. All these factors require a change in regular driving behaviour.

As noted, it is crucial that drivers are aware that they are distracted. If they are not aware then they will not engage in behavioural adaptation; and even though they may not experience an increase in workload, safety margins will be further eroded. In this respect an important study was conducted by Lesch and Hancock (2004) of whether drivers are aware that the use of cellphones (mobiles) during driving may result in associated performance decrements. The study showed that some groups (women and in particular older women) express disproportionately high confidence relative to their actual performance in the presence of the cellphone task. In other words, it seems that using a cellphone distracted older women drivers quite a lot, even though they were not aware of it. This is important because for drivers who do not realize they are distracted, no compensatory strategies will be engaged. Obviously this may result in dangerous behaviour even though these drivers do not experience an increase in workload.

The most dangerous situations may occur when unexpected events happen to which the driver has no immediate response. Especially when such events cannot be anticipated (a quiet motorway drive and suddenly there is stationary traffic) drivers may experience extreme workload and possibly panic reactions may occur (for example, they decelerate severely and/or make erratic manoeuvres). Even though these incidences cannot always be anticipated (for example a tyre blowout), it is crucial that in some way drivers are warned that such an event may occur.

Conditions that Increase or Decrease Driving Task Workload

Table 5.1 gives an exhaustive overview of conditions that may affect workload levels. The following section examines some of these factors in greater detail and discusses some recent findings. In most cases workload increase or decrease is the result of interactions of those factors outlined in Table 5.1.

Increase in Workload

Road environment The field study by Verwey (2000) discussed earlier that involved an experiment with an instrumented vehicle showed that increases in workload are typically associated with visual aspects of the driving task, especially when it involves subtasks at the manoeuvring level. A recent study by Cantin et al. (2009) using a driving simulator also showed an increase in workload as the complexity of the driving context increased. Crucially, relative to young drivers, older drivers showed a disproportionate increase in workload with increasing complexity of the driving environment. As in Verwey (2000) driving at intersections and overtaking manoeuvres required more mental workload than straight road driving. These findings, suggesting that older drivers may experience a high level of load at complex intersections, fits in with the findings showing that most accidents involving older drivers occur at intersections (for example, Hakamies-Blomqvist et al. 1999). A study by Hancock, Wulf and Thom (1990) confirms these findings. Their study reported increased workload (as measured by the above-mentioned subjective workload measures) and head movements during turn sequences (for example, left turns at crossings) compared to straight road driving. Hancock et al. (1990) pointed out that higher workload may be associated with failure to detect oncoming cars. In another recent study carried out on a closed circuit, Makishita and Matsunaga (2008) demonstrated that, compared with young drivers, elderly drivers experience a much higher workload as measured by a secondary auditory reaction time task. Importantly, however, there were large individual differences, indicating that even some young drivers showed signs of workload overload.

Car environment The car environment has many potential distractions that may increase workload substantially. Several studies have shown that talking on the phone while driving significantly increases driver workload and slows reactions (Alm and Nillson 1995). Some have argued that talking on the phone is just as bad as being drunk behind the wheel (Strayer, Drews and Crouch 2003). There is also evidence that drivers talking on the phone slow down substantially in order to keep their overall workload at an acceptable level (Haigney and Westerman 2001; Strayer et al. 2003). Even though in some countries handheld phones are not allowed in the car, research shows that there is basically no difference between hand-held and hands-free phones, both having similar effects in terms of driver performance (Lamble et al. 1999) and mental workload (Törnros and Bolling 2005). The underlying notion is that talking on the phone while driving increases workload regardless of whether the phone is hand-held or hands-free.

Recently, in a simulator study Drews et al. (2009) investigated the effect of text messaging during driving. This study showed that drivers who were texting responded more slowly to the onset of brake lights of the car in front and showed impairment in lateral and forward control compared to drivers who were not texting. The authors concluded that text messaging while driving results in a dramatic increase in workload, even exceeding the impact of a phone conversation while driving.

Decrease in Workload

Navigation systems Even though some in-car systems may prove to cause an increase in workload there are also various systems that may help reduce workload. For example a study conducted in the Netherlands showed that drivers with a navigation system required less mental effort than those without a system. This was based on subjective ratings as well as objective measurements (Vonk et al. 2007). Moreover, this study showed that drivers with a navigation system covered shorter distances (-16 per cent) and therefore spent less time in traffic (-18 per cent). Perez et al. (1996) reports accident data covering a year in which one group of vehicles had a navigation system and one group did not. They reported that the use of a navigation system has a favourable effect on driving performance. A small positive (yet not significant) effect was also reported on the number of incidents, crashes and near-crashes. Note, however, that out-of-date or incorrect information in the navigation system can lead to great confusion, which in turn may result in severe stress and increased workload. From a workload point of view, adjusting the navigation system while driving also increases workload as less attention can be devoted to the driving task.

Driver assistance systems A recent driving simulator study by Davidse et al. (2009) showed that an advanced driver assistance system (ADAS) helped older drivers navigate though urban areas. The ADAS can provide messages informing drivers about right-of-way regulations, obstructed views and about safe gaps when joining or crossing traffic streams. Even though the system did not reduce the workload (for which it was originally designed) it did improve driving performance for all age groups.

A recent study by Brookhuis et al. (2009) investigated the effect of a so-called ‘Congestion Assistant’, a system that combined features of a congestion warning system (acoustic warning and accelerator pedal counterforce) and a stop and go system (automatic accelerator and brake during congestion). In this driving simulator study workload was measured by means of heart rate, a secondary task and by subjective scaling techniques. The results showed that driving with the Congestion Assistant while in congestion potentially leads to reduced driver mental workload, while just before the congestion started the system may have increased the workload. Overall, the participants appreciated the Congestion Assistant.

Summary and Conclusions

It is clear that driver workload is an important factor influencing traffic behaviour. With increasing demands for attention from in-car systems as well as from relevant and irrelevant information along the roads, it is important that workload remains within acceptable limits. Because drivers engage in compensation strategies when overload becomes imminent, some argue that workload in driving is never really a problem. However, even though most of the time these strategies are adequate, it should be noted that with every distraction safety margins are eroded. It is the addition of extra, unexpected distractions on a driver who is already compensating for high workload that ultimately may lead to collisions.

References

Alm, H. and Nilsson, L. 1995. The effects of a mobile telephone task on driver behaviour in a car following situation. *Accident Analysis and Prevention*, 27(5): 707–15.

Cantin, V., Lavalliere, M., Simoneau, M. and Teasdale, N. 2009. Mental workload when driving in a simulator: Effects of age and driving complexity. *Accident Analysis and Prevention*, 41(4): 763–71.

Brookhuis, K.A., van Driel, C., Hof, T., van Arem, B. and Hoedemaeker, M. 2009. Driving with a congestion assistant: Mental workload and acceptance. *Applied Ergonomics*, 40(6): 1019–25.

Drews, F.A., Yazdani, H., Godfrey, C.N., Cooper, J.M. and Strayer, D.L. 2009. Text messaging during simulated driving. *Human Factors*, 51(5): 762–70.

Davidse, R.J., Hagenzieker, M.P., Wolffelaar, P.C. van, and Brouwer, W.H. 2009. Effects of in-car support on mental workload and driving performance of older drivers. *Human Factors*, 51: 463–76.

De Waard, D. 1996. *The Measurement of Drivers' Mental Workload*. PhD thesis, University of Groningen. Haren: University of Groningen, Traffic Research Centre.

Gopher, D. and Donchin, E. 1986. Workload – an examination of the concept. In: K. Boff, L. Kaufman and J. Thomas (eds), *Handbook of Perception and Performance: Vol. 2 – Cognitive Processes and Performance*. New York: Wiley, 41.1–41.49.

Haigney, D. and Westerman, S.J. 2001. Mobile (cellular) phone use and driving: A critical review of research methodology. *Ergonomics*, 44: 132–43.

Hancock, F.P., Wulf, G. and Thom, D. 1990. Driver workload during differing driving maneuvers. *Accident Analysis and Prevention*, 22(3): 281–90.

Hancock, P.A., Lesch, M. and Simmons, L. 2003. The distraction effects of phone use during a crucial driving maneuver. *Accident Analysis and Prevention*, 35(4): 501–14.

Hakamies-Blomqvist, L., Mynttinen, S., Backman, M. and Mikkonen, V. 1999. Age-related differences in driving: Are older drivers more serial? *International Journal of Behavioral Development*, 23: 575–89.

Kahneman, D. 1973. *Attention and Effort*. Englewood Cliffs, NJ: Prentice Hall.

Lamble, D., Kauranen, T., Laakso, M. and Summala, H. 1999. Cognitive load and detection thresholds in car following situations: Safety implications for using mobile (cellular) telephones while driving. *Accident Analysis and Prevention*, 31(6): 617–23.

Lesch, F. and Hancock, P.A. 2004. Driving performance during concurrent cellphone use: Are drivers aware of their performance decrements? *Accident Analysis and Prevention*, 36(3): 471–80.

Makishita, H. and Matsunaga, K., 2008. Differences of drivers' reaction times according to age and mental workload. *Accident Analysis and Prevention*, 40(2): 567–75.

Michon, J.A. 1985. A critical view of driver behavior models: What do we know, what should we do? In: L. Evans and R. Schwing (eds), *Human Behavior and Traffic Safety*. New York: Plenum Press, 485–520.

Perez, W.A., van Aerde, M., Rakha, H. and Robinson, M. 1996. *TravTek Evaluation Safety Study* (FHWA-RD-95-188). Washington, DC: Turner-Fairbank Highway Research Center, US Department of Transportation, Federal Highway Administration.

Strayer, D.A., Drews, F.A. and Crouch, D.J. 2003. Fatal distraction? A comparison of the cell-phone driver and the drunk driver. In: D.V. McGehee, J.D. Lee, M. Rizzo, M. Raby and L. Boyle (eds), *Proceedings of the 2nd International Driving Symposium on Human Factors in Driver Assessment, Training and Vehicle Design*. Iowa City, IA: University of Iowa, 25–30.

Theeuwes, J. and Godthelp, H. 1995. Self-explaining roads. *Safety Science*, 19: 217–25.

Törnros, J.E.B. and Bolling, A.K. 2005. Mobile phone use – effects of handheld and handsfree phones on driving performance. *Accident Analysis and Prevention*, 37(5): 902–9.

Van Winsum, W., Martens, M. and Herland, L. 1999. The effects of speech versus tactile driver support messages on workload, driver behavior and user acceptance (Report TM-99-C043). Soesterberg: TNO Human Factors Research Institute.

Veltman, J.A. and Jansen, C. 2003. Differentiation of mental effort measures: Consequences for adaptive automation. In: G.R.J. Hockey, A.W.K. Gaillard and O. Burov (eds), *Operator Functional State: The Assessment and Prediction of Human Performance Degradation in Complex Tasks*. Amsterdam: IOS Press, 249–59.

Verwey, W.B. 2000. On-line driver workload estimation: Effects of road situation and age on secondary task measures. *Ergonomics*, 43: 187–209.

Vonk, T., Rooijen, T. van, Hogema, J. and Feenstra, P. 2007. Do navigation systems improve traffic safety? (TNO Report 2007-D-R0048/B). Soesterberg: TNO Mobility and Logistics.

Wickens, C.D. 1984. Processing resources in attention. In: R. Parasuraman, J. Beatty and R. Davies (eds), *Varieties of Attention*. New York: Wiley, 63–101.

Wickens, C.D. and Kessel, C. 1981. Failure detection in dynamic systems. In: J. Rasmussen and W.B. Rouse (eds), *Human Detection and Diagnosis of System Failures*. New York: Plenum Press.

Young, M.S. and Stanton, N.A. 1997. Automotive automation: Investigating the impact on drivers' mental workload. *International Journal of Cognitive Ergonomics*, 1(4): 325–36.

Chapter 6
Information Carriers Meet Basic Ergonomic Principles

Richard van der Horst

Individual information carriers should meet basic ergonomic principles to be visible, clear and understandable for the road user.

Theoretical Background

Road users must read and comprehend from the road environment those aspects that the designers intended to communicate to them (Campbell, Richard and Graham 2008). What road users may comprehend and find totally logical may not be the same as what the designers had in mind. The way people perceive and collect information from the environment depends strongly on some typical sensory (mainly visual) characteristics on the one hand and characteristics of the information carriers on the other. For example, to be able to detect something in the environment it is important to take into account human sensory characteristics with respect to light intensity, contrast (in brightness and colour) and conspicuity of objects. Following the detection of information (with the precondition: Is it visible?), the next stage in information processing deals with cognitive functions for which attention, 'expectancy', memory, recognition and comprehension are relevant factors. The steps in the information processing sequence take time, and for communicating properly and timely with the road user several requirements for information carriers can be derived from basic ergonomic principles.

Some Mechanisms on the Human Sensor Input Side

Visual perception is crucial for conducting the driving task. Estimates are that over 90 per cent of information needed for driving is visual and to a much smaller extent auditory (Hills 1980). Also the equilibrium system may be an important source in cases of high longitudinal or lateral acceleration or vibrations to inform and warn the driver of potentially unsafe conditions. All senses have a number of elementary characteristics in common:

- They all have an *absolute threshold*: The amount of incoming energy in the senses (light, sound) must be of a given minimum level to be perceivable at all.

- Senses display *adaptation*: The longer a stimulus of a given constant intensity is presented to a sense, the less reaction of that sense is to be expected. Adaptation of the eyes to light or dark levels (as in a cinema) is a well-known daily experience. As Boff and Lincoln (1988) state: 'When stimulated by light, the visual system progressively loses sensitivity. In the dark, sensitivity recovers, a process known as dark adaptation'.
- *Differential thresholds*: The difference between two stimuli (for example two sounds or colours) must be at a given level to be perceived. The threshold value for this difference is roughly a constant fraction (the so-called 'Weber fraction') of the stimulus against which the difference has to be detected. For example, if one weighs 100 coins by hand, the addition of one coin will not be noticed. Experimentally, one can determine that at least five coins should be added to notice the difference with 100 coins. However, if one starts with 1,000 coins then 50 coins should be added to feel the difference. The Weber fraction for weight, the just noticeable relative difference, is about 5 per cent. These fractions differ among the senses.
- *Non-linearity*: The relationship between a stimulus in physical units expressed and the resulting perceived sensation is non-linear. The perception of electrical current is an extreme example: doubling the micro-amperes presented to a fingertip results in a perceived experience four times as strong. Contrary to this, objectively increasing the light level of a lamp by a factor of 2 results in a perceived increase in experience of 0.25.

In order to communicate a message to road users it is important to take these sensory characteristics into account with respect to light intensity, contrast and conspicuity of objects. Engel (1976) gave the following definition of conspicuity: '*Visual conspicuity* is the degree of perceptual prominence of a visible object in its surroundings by virtue of crude sensory features such as differences in brightness, colour, outline, size, movement, etc.' – the degree to which an object is intrinsically attracting attention from a human observer. Visual conspicuity can be expressed as the maximum eccentricity angle (relative to the forward field of view direction) at which an object can be detected. The larger this angle, the more conspicuous the object will be. Wertheim (1986) describes an optical method that enables a quick and reliable real-world measurement of conspicuity of a given element in its surroundings.

Visual acuity (visus) is another aspect that is relevant for driving. It is the ability to perceive small details and can be measured by various reading cards, of which the Landolt-C is most common in the Netherlands. This test measures the visual angle of the smallest perceivable detail (operationalized as a gap in a black circle on a white background) and visual acuity is expressed as the reciprocal value of this angle. In ideal circumstances, the smallest detail that can be distinguished equals about 1 minute of arc, or 1 mm at a distance of 3.5 m. The visual acuity of Dutch road users aged 18 and above has a normal distribution, with an average

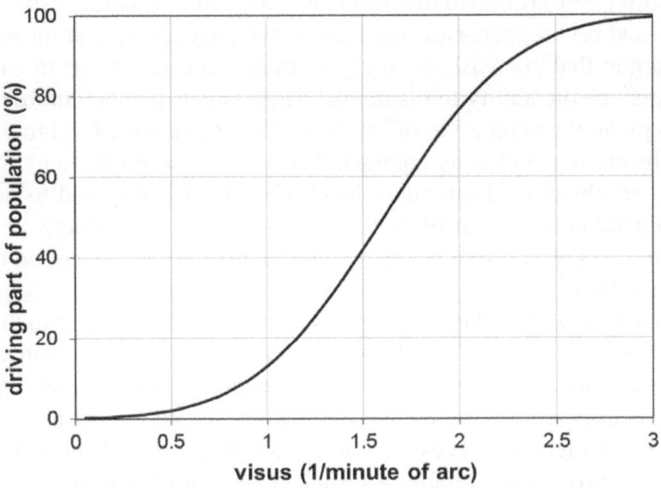

Figure 6.1 Distribution of visual acuity of Dutch road users

of 1.61 and a standard deviation of 0.54 (van Norren 1981; Janssen, Lourens and Göbel 1998) – see Figure 6.1. For design purposes, usually an 85th percentile value is taken to ensure that the majority of the population meets this requirement. As is clear from Figure 6.1, 85 per cent of the driving population has a visus higher than about 1.

Such a value is a good basis for defining required legibility distances. Despite the fact that the relevance of *static* visual acuity to safe driving is hardly ever questioned, for a period of over 30 years more than 100 independent studies failed to find a significant relationship between low static visual acuity and crash involvement (Burg 1968; Shinar 1977); and more recent research still suggests that if there is a relationship, the evidence is quite weak (Shinar 2007). Shinar (2007) gives several reasons why the lack of empirical association between visual acuity and crash involvement exists. These include age effects – the high crash risk group of young drivers having the best visual acuity and older drivers with the worst visual performance being cautious on the road both in the way they drive and in selecting the time and place to drive; and a 'restriction to range' effect – all studies were conducted on licensed drivers who are expected to have (corrected) visual acuity that falls within the requirements for driving. Moreover, most objects we have to perceive to drive safely, such as cars or pedestrians, are relatively large.

Dynamic visual acuity, a measure of acuity when we are in relative motion to another object, seems to be much more relevant to a driving environment. Burg (1968) and Shinar (1977) indeed found that dynamic visual acuity related stronger to the occurrence of a crash than static visual acuity, although the relationship was still weak. Dynamic acuity may be especially relevant in relation to aging. A study by Burg (1966) revealed that whereas static acuity is hardly affected by age, dynamic visual acuity decreases significantly with age, starting at the age of 40.

Some argue that *contrast sensitivity* is even more important to safe driving than acuity, since the ability to distinguish large targets is much more relevant to safe driving than the perception of small details. The relative brightness between adjacent objects is denoted as contrast. No matter how much light is available or how big an object is, a minimum level of contrast is required to perceive an object. Reduced contrast can be experienced when driving directly into the sun or in darkness. As with visual acuity, contrast sensitivity is reduced in the elderly (Kline et al. 1990).

Low illumination and glare can create specific problems related to the adaptive characteristics of the eyes. When the amount of available light suddenly increases (for example glare from the headlamps of oncoming cars at night, or driving into sunlight) or decreases (when entering a tunnel or after a car has passed with headlights on at night) our eyes have to (re)adapt to these changed light levels. The colour-sensitive cones require a sufficient amount of light while the rods are inactive with high levels of illumination. This adaptation process in transitions from light to dark and vice versa takes time. It is important to note that adapting to a dark environment from a bright environment takes much longer than vice versa. For dark adaptation reaching the maximum sensitivity of the cones takes about 8 minutes, whereas the rods need up to 20 minutes to adapt. Adaptation to bright light is completed within at most a few seconds. For this reason tunnel entrances are more critical than tunnel exits. Acuity at low light levels and in the presence of glare is significantly lower than in optimal light conditions. Moreover, night-time acuity and acuity under glare significantly decrease after the age of 60 (Shinar 2007). Theeuwes, Alferdinck and Perel (2002) studied the effects of discomfort glare (the subjective sensation of discomfort as experienced due to glare) on driving behaviour. The results showed that a relatively low glare source caused a significant drop in detecting (simulated) pedestrians along the roadside and made participants drive slower on dark and winding roads, with the largest effects for older drivers.

Whereas distinct measures of the 'passive' visual system seem to show weak to moderate relations to accident probability, the so-called 'Useful Field of View' (UFOV) is more clearly related to crash rates. According to Shinar (2007), the UFOV is a composed measure of visual information processing speed:

1. without distraction (distinguishing objects in the central field of view);
2. with divided attention (a peripheral target has to be detected); and
3. with selective attention (detection of peripheral target with cluttered visual field).

These tasks do correlate with measures of visual ability (such as acuity and contrast sensitivity) but they also relate to cognitive abilities. Ball and Owsley (1993) showed that the UFOV significantly distinguishes between crash-free and crash-involved drivers. This reflects the importance of both mental processing abilities and visual abilities in safe driving.

Some Consequences for Driving

As illustrated in the previous section, glare may induce behavioural adaptation effects of reducing speed (with consequences for traffic performance) and cause poor detection of roadside pedestrians (safety issue). In curves on motorways, anti-glare screens on the guardrails in the median are a well-known feature to prevent glare from the headlamps of oncoming traffic. Without public lighting, pedestrians are being detected poorly at night, and only wearing retro reflective material may be effective.

The perception of motion and velocity in traffic is a good example of how the absolute and differential threshold of our eyes may work out. The absolute threshold value for perception of an object is about 2 minutes of arc/s. Because the threshold for movement is expressed as an angular value, the object can be nearby and move slowly or be far away and move fast. Without further depth cues the eyes are not able to make the proper distinction. The consequence of this is illustrated in Table 6.1, which gives just noticeable speed differences with a lead vehicle (as measured with test persons) at given following distances (Harvey and Michon 1971).

Doubling of the observation distance would require a more than twice as large speed difference to be detected. Expressed in time units, it would result in a more risky situation if one is following at a greater distance (of course given the assumption that one only looks for 1 s). The eye that is functioning on the basis of visual angle and angular velocity cannot do much about it, as the eye with a fixed sensitivity for motion works less at a greater distance.

The adaptation issue also plays a role in motion and speed perception. For example, habituation of one's own speed while driving at high speed on a motorway may result in a much lower speed perception when entering a lower-order road

Table 6.1 Just noticeable speed difference with a lead vehicle (measured in the dark, based only on relative movement of the rear lights of the lead vehicle for dark-adapted test persons and a viewing time of 1 s)

Observation distance (in metres)	Just noticeable speed difference (in km/h)
40	5
80	11
160	27
320	64
640	151

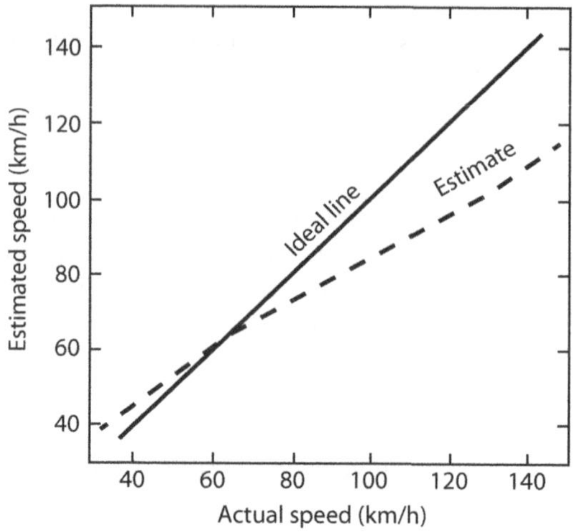

Figure 6.2 Relationship between physical and estimated speed

than actually is the case (Denton 1966). One may then have the feeling that one is driving much too slowly and increase speed. The adaptation process to a high speed will be finished in a short period of time (1–2 minutes on a motorway), whereas the opposite takes much longer.

The non-linearity between objectively and subjectively perceived speed is illustrated in Figure 6.2. Physical and perceived speed differ from each other: up to about 65 km/h there is an overestimation of perceived speed; above that level underestimation is the case (van der Horst and Janssen 2005). Recarte and Nunes (1996) showed that such an underestimation is also present in speed production. When people are asked to adjust a vehicle's speed (60, 80, 100 or 120 km/h) to a predefined value there is a tendency to underestimate speed, which results in overcompensation in the active production task. This effect was more pronounced at lower speeds.

In addition to the sensory characteristics relevant for speed perception as mentioned above, eye-height above ground level also determines the subjective speed experience. A rule of thumb is that speed impression equals roughly the number of eye-heights one is travelling (in m/s). A pedestrian is walking in a forward direction at a speed of about 1 eye-height/s; a driver in a passenger car (eye-height of 1.2 metres) gets a speed impression of about 10 (at 50 km/h); and a truck driver (eye-height of 2.5 metres) gets an impression of about 5. This is exactly why a person in a truck thinks they are driving much slower than in a passenger car at the same speed. Evidence suggests that drivers of sport utility vehicles (SUVs) and minivans drive faster than drivers of regular passenger cars when asked to drive at a motorway speed at which they feel comfortable and safe (Rudin-Brown 2004). In the ultimate case of sitting in an aeroplane the ratio is about 0.03, which is why it looks as if one is moving rather slowly relative to the ground surface.

Table 6.2 Examples of task level by information type, carrier, processing stage and potential actions (after Godthelp and Tenkink 1990)

Task level	**Information type**	**Information carrier**	**Information processing**	**Potential actions**
Navigation	Route	Route sign	Read	Turn, exit
	Rules	Traffic signs, signals	Detect, recognize	Slow down, stop
Guidance	Obstacles	Illumination, reflectors	Detect, localize	Swerve, stop
	Other traffic	Vehicle signalling	Detect, estimate path and speed	Overtake, slow down, cross
Control	Road alignment	Beaconing	Recognize	Adjust speed
	Own position	Road markings	Estimate own path and speed	Steer

Driving Task Related Issues

As indicated in Chapter 4, the three hierarchical levels of driving tasks, as well as the different stages in human information processing, should be taken into account in roadway design. Godthelp and Tenkink (1990) give a good overview of examples that relate these two dimensions to each other (see Table 6.2). In general, in the perception process successive steps of human information processing reflect characteristics of the information source such as *detectability*, *recognizability*, *legibility* and *comprehensibility*.

- *Detectability* includes aspects such as how visible and how conspicuous the information is relative to its context (localization, surroundings etc.).
- *Recognizability* refers to how well acquainted the road user is with the appearance of the information carrier and whether the user has previous experience of it (or it directly connects with their knowledge).
- *Legibility* refers to a road user's ability to see or read the specific details of the information (either text or pictogram).
- *Comprehensibility* refers to the process whereby the road user understands the meaning of the message and elicits the correct behaviour (or not, if they intentionally decide not to obey it).

Elderly Road Users

Older road users may experience specific problems in processing information in the driving environment. Lower workload capacity and a reduced peripheral field of vision lead, for example, to problems in detecting and estimating the speed of other traffic. On the other hand, people are living longer and therefore want to stay mobile for a longer time. The proportion of the elderly in the population will increase considerably. Not only will the share of older people increase, they will also be more active, more often have a driving licence and participate more in traffic (Whelan et al. 2006) Therefore, the approach that the environment has to be 'designed for all', taking into account problems older drivers may experience, is gaining more and more acceptance. Specific recommendations may apply to the roadway as well as the car. Apart from a general finding that older people need more time for processing information (resulting in longer reaction times), they also experience more problems with orientation, which might induce stress and distraction from the driving task when looking for the correct route. In-car navigational systems can help reduce these problems. On the other hand, operation of these systems can also induce cognitive and visual distraction. It is therefore important to adjust these systems (for example by a heads-up display or a simplified display) in such a way that they meet the capacities of older drivers (Wilschut 2009). A good example of this is the total voice-based driver aid and navigation system of General Motors. With one push of a button, the driver is connected to an operator who can assist with tasks from route guidance to directory assistance. Environmental design considerations could include, for example, increasing overhead road illumination, simplifying intersections and road widening. Whereas it is generally assumed that such adjustments benefit all road users, this may not always be the case. Wider roads can improve lane keeping but can also provoke faster driving. For example, Elvik and Vaa (2004) found that increasing the width of the road reduced the number of crashes on roads in rural areas, but may lead to a small increase in the number of crashes in urban areas. The results of increasing the width of traffic lanes (in a range of 0.3–0.5 metres) appeared to be inconsistent with respect to road safety consequences.

Functional limitations will, eventually, develop with age. Davidse (2007) gives a good overview of these sensory, cognitive, decision making and executive limitations. Elderly persons have, on average, poorer sight; find it more difficult to filter relevant information and take the appropriate decisions; and their reaction time is longer compared to younger persons. These limitations will affect driving performance (Wood et al. 2006), but do not lead automatically to a higher accident risk because many older drivers tend to take compensatory measures (Hakamies-Blomqvist and Peters 2000). At intersections, the three hierarchical levels of the driving task are often competing for cognitive attention from the driver, and older drivers increasingly experience negotiating an intersection as a loading task. For example, older drivers require gaps that are approximately 1 second longer and tend to reject more usable gaps than younger drivers (Lerner et al. 1995). Campbell et al. (2008) discuss the problems older drivers especially may have with skewed intersections because of the drivers' limitations in body flexibility. From a human factors point of view, skewed intersections should be avoided as much as possible;

perpendicular intersections are preferable. At signalized intersections the driving task is substantially simplified compared to non-signalized ones because at the guidance level of the driving task the interaction with crossing traffic is separated in time. For older drivers, the left-turn manoeuvre appears to be more difficult than for younger drivers. Staplin et al. (2001) indicate that older drivers are over-involved in these manoeuvres. An exclusive left-turn phase, in particular for situations with two lanes of opposing traffic, is common in the Netherlands for most signalized intersections.

In road systems design it is important to take limitations and abilities of all road users into account (see also Chapter 10).

Resulting Design Requirements

As all information processes take time, it is crucial that road users have the opportunity to conduct all necessary steps. Tasks that develop such that they can be conducted at the skill-based level of task performance are the most efficient and take the least time for the road user to act upon (see Chapter 4). Driving can be regarded as a time-sharing task and how to design a roadway and traffic management environment will strongly depend on the speed that the environment induces. Therefore, the design speed – the speed with which the designer calculates various design elements from the point of view of the intended road users – is crucial.

Relevant information for the driving task has to be detected first of all. Detection is defined by object conspicuity (the extent to which an object is prominent in the scene), depending on factors such as object size, contrast and location. Detection distance (somewhat synonymous with visibility distance) is the distance at which a representative group of the population (for example the 85th percentile) can detect or see an information carrier. Of course, this distance will have to fall within the area defined by the 'line of sight' of the road at hand. A visually conspicuous object does not, however, necessarily attract attention when it is not relevant to the task at hand. For navigation, the uniformity and continuity of route signing devices are important issues. Uniformity in design, layout and placement will ensure that route signs are recognized (or identified) as such rather quickly (it is not without reason that commercial advertisements along roads sometimes try to copy the layout of route signs).

A timely view of overhead route signs is an important aspect of good route signing. Figure 6.3 gives an example of the problem: a viaduct obscures the view on a route sign and the road user can only make the link between the route sign directions and the proper path choice at a rather late stage. By painting the road numbers on the road surface, the road administrator has tried to reduce this problem that the road user encounters at this particular motorway section.

The next step in information processing is legibility: the information on a sign should be readable for the driver for a long enough period of time. In the literature,

Figure 6.3 Example of poor visibility of a route sign at the decision point due to the presence of a viaduct at a motorway diversion section

based upon British research by Odescalchi, Rutley and Christie (1962), for a long time the following reading time formula was used (van Norren 1981):

$$T = N/3 + 2 \text{ (in seconds)}$$

With:

T = necessary reading time
N = number of information units on the sign

Based upon a given driving speed (that determines the available time to read the sign by a driver), one can calculate how many names should be allowed on the sign. This would result in a limited number of five or six at most. Alblas and Janssen

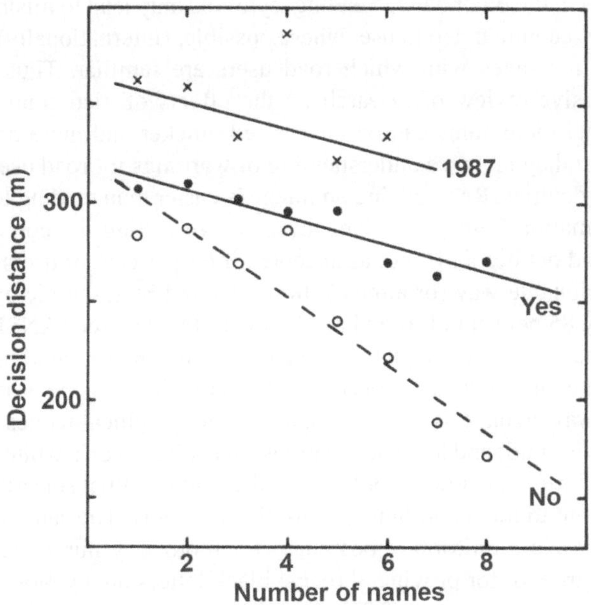

Figure 6.4 Decision distance as a function of the number of names on a route sign (Alblas, Janssen and Buist 1989)

Note: Yes-response results from a previous study (Alblas and Janssen 1987) are given as well.

(1987) and Alblas, Janssen and Buist (1989) investigated necessary reading times under the assumption that a driver searching for a given destination does not need to read all the names, just in case the right name is on the route sign. On average, they have to check half the names to find the one they are looking for. The results of this research in a field experiment with test persons as passengers in a vehicle driving along 190 km of unfamiliar motorways are given in Figure 6.4. The Yes-responses refer to the situation where the target name was present on the sign, whereas the No-responses reflect the situation where the target name was absent. The more names present, the longer it takes to decide – and even more when the target name is not present on the route sign.

From this research, it was concluded that providing eight names on a route sign would still be acceptable. Extrapolation would suggest that even ten names would be workable on motorways since overhead signs disappear from the line of sight at about 50 metres before the gantry. A similar value holds for signs that are located at the side of the road.

In general, providing drivers with detailed written messages is not practical due to processing time and restricted sign sizes. Where applicable, symbols (icons) or pictograms are preferred over text, since they are faster to read and comprehend. On the other hand, 'reading' symbols may lead to misinterpretations. It is therefore recommended to use, where possible, (internationally) standardized symbols and messages with which road users are familiar. Tijus et al. (2007) give an extensive review of research on the effects of pictograms, resulting in the finding that pictograms can be interpreted quicker and more accurately than words and that they improve understanding of warnings for road users with visual or literacy difficulties. Recognizing an image is easier than reading text. ISO 9186 (Public Information Signs) describes a method of testing the comprehension of pictograms and defines a symbol as accepted if 67 per cent of users understand it in an unquestionable way (or almost). In the United States, a pictogram must be understood by 85 per cent of users in order to be standardized (ANSI Z535-1987).

A combination of word picture and distinctive representation of different destination categories may also speed up information processing. For example, the Dutch motorway signage system distinguishes geographical names (white letters on a blue background) and local destinations (black letters on a white background). So, in case of a mix of names, people can make a quick initial selection of items on the sign relevant to them and ignore non-relevant items. The same holds for road numbers that have a uniform appearance (for motorway numbers: white letters on a red background; for provincial roads: black letters on a yellow background). Figure 6.5 gives an example of the Dutch route signage system with overhead signs on a motorway.

Figure 6.5 Example of Dutch motorway route signage with geographical and local destinations and road numbers distinguished by different layouts

Letter and symbol sizes should be such that road users can actually read them in time while driving at a given design speed. The legibility coefficient is expressed in metres per centimetre (m/cm) of letter height. The Dutch ANWB E-font alphabet has a coefficient of 6.2 m/cm, implying that letter heights on motorway signs with a legibility distance of 220 metres should be 36 and 27 centimetres for capitals and lowercase letters, respectively (Walraven, Varkevisser and Bijl 1996). For text on dynamic signs, in daytime conditions the brightness of the letters/symbols needs to compete with sunshine, whereas in night-time conditions the brightness of the letters/symbols needs to be dimmed considerably to avoid overexposure and glare.

A sign can be easily detectable in its surroundings, recognizable and legible, but still not be comprehended, meaning that road users do not give the message the right meaning. For example, in Figure 6.5, the indications 'Ring Rotterdam (noord)' and 'Ring Rotterdam (oost)' match the geographical locations North ('noord') and 'East ('oost') of the beltway ('ring'). However, in the case of the beltway of Amsterdam the indication 'beltway Amsterdam East' refers to the southern part of the beltway in a westerly direction as opposed to the other direction indicated by 'beltway Amsterdam North' going east on the southern part of the beltway, completely misunderstood by many road users. The mental map road users may have of the network does not match the information provided.

Summary and Conclusions

In this chapter characteristics of individual information carriers are confronted with some elementary characteristics of the human senses. Road users must be able to perceive the roadway and understand what the motorway designer and traffic engineer intended to communicate. The chapter describes the way people perceive and collect relevant information from the environment that has some consequences for driving and driving task related issues. As the proportion of older people in the driving population will increase, some specific problems elderly drivers may encounter in perceiving and interpreting information are summarized. Some implications for designing road systems are discussed. Information carriers should be timely visible (detectable in their surroundings), recognizable and legible in the first instance, and be comprehended by the majority of road users in an unambiguous manner.

References

Alblas, B.P. and Janssen, W.H. 1987. *Een veldstudie over de informatiehoeveelheid op wegwijzers* [*A Field Study on the Amount of Information on Traffic Signs*] (TNO Report IZF 1987 C-31). Soesterberg: TNO Instituut voor Zintuigfysiologie (in Dutch).

Alblas, B.P., Janssen, W.H. and Buist, M. 1989. *Een veldstudie over de informatiehoeveelheid op wegwijzers II* [*A Field Study on the Amount of Information on Traffic Signs II*] (TNO Report IZF 1989 C-14). Soesterberg: TNO Instituut voor Zintuigfysiologie (in Dutch).

Ball, K. and Owsley, C. 1993. The useful field of view test: A new technique for evaluating age related declines in visual function. *Journal of the American Optometric Association*, 64: 71–80.

Boff, K.R. and Lincoln, J.E. 1988. *Engineering Data Compendium: Human Perception and Performance*. Wright-Patterson Air Force Base, OH: Armstrong Aerospace Medical Research Laboratory (AAMRL).

Burg, A. 1966. Visual acuity as measured by dynamic and static tests: A comparative evaluation. *Journal of Applied Psychology*, 50(6): 460–66.

Burg, A. 1968. Lateral visual field as related to age and sex. *Journal of Applied Psychology*, 52: 10–15.

Campbell, J.L., Richard, C.M. and J. Graham 2008. *Human Factors Guidelines for Road Systems: Collection A – Chapters, 1, 2, 3, 4, 5, 10, 11, 13, 22, 23, 26* (NCHRP Report 600A). Washington, DC: Transportation Research Board (National Cooperative Highway Research Program).

Davidse, R.J. 2007. *Assisting the Older Driver: Intersection Design and In-Car Devices to Improve the Safety of the Older Driver*. PhD thesis. Leidschendam: SWOV Institute for Road Safety Research.

Denton, G.G. 1966. A subjective scale of speed when driving a motor vehicle. *Ergonomics*, 9: 203–210.

Elvik, R. and Vaa, T. 2004. *The Handbook of Road Safety Measures*. Amsterdam: Elsevier.

Engel, F.L. 1976. *Visual Conspicuity as an External Determinant of Eye Movements and Selective Attention*. PhD thesis, Technical University of Eindhoven.

Godthelp, J. and Tenkink, E. 1990. *Zichtcriteria voor wegen en informatiedragers langs de weg* [*Sight Distance Criteria for Road Design and Roadside Information*] (TNO Report IZF 1990 C-10). Soesterberg: TNO Instituut voor Zintuigfysiologie (in Dutch).

Harvey, L.O. and Michon, J.A. 1971. *The Perception of Manoeuvres of Moving Vehicles. Progress Report I: Effects of Viewing Distance and Angular Separation* (TNO Report 1971-C6). Soesterberg: TNO Institute for Perception.

Hakamies-Blomqvist, L. and Peters, B. 2000. Recent European research on older drivers. *Accident Analysis and Prevention*, 32(4): 601–7.

Hills, B.L. 1980. Vision, visibility, and perception in driving. *Perception*, 9: 183–216.

Horst, A.R.A. van der and Janssen, W.H. 2005. De weggebruiker als informatieverwerkend systeem [The road user as an information processing system]. In: *Syllabus PAO-cursus Ontwerpen van Wegen en Kruispunten, OWK1, 8, 9, 15, 16 November 2005*. Delft: Stichting Postacademisch Onderwijs Vervoerswetenschappen en Verkeerskunde (in Dutch).

Janssen, W.H., Lourens, P.F. and Göbel, M.P. 1998. *Kenmerken van weggebruikers en hun relatie met gedrag en veiligheid* [*Road User Characteristics and their Relation with Behaviour and Safety*] (TNO Report TM-98-C042). Soesterberg: TNO Technische Menskunde (in Dutch).

Kline, T.J.B., Ghali, L.M., Kline, D.W. and Brown, S. 1990. Visibility distance of highway signs among young, middle-aged and older observers: Icons are better than text. *Human Factors*, 32: 609–19.

Lerner, N., Huey, R.W., McGee, H.W. and Sullivan, A. 1995. *Older Driver Perception-Reaction Time for Intersection Sight Distance and Object Detection: Volume 1. Final Report* (Report FHWA-RD-93-168). Washington, DC: Federal Highway Administration.

Norren, D. van 1981. *Informatiedragers langs de weg: een overzicht van zichtbaarheidsproblemen* [*Information Carriers along the Road: A Survey of Visibility Problems*] (TNO Report IZF 1981 C-25). Soesterberg: TNO Instituut voor Zintuigfysiologie (in Dutch).

Odescalchi, P., Rutley, K.S. and Christie, A.W. 1962. *The Time Taken to Read a Traffic Sign and Its Effect on the Size of the Lettering Necessary* (Note No. LN/98/PO.KSR.AWC). Crowthorne: Road Research Laboratory (unpublished report).

Recarte, M.A. and Nunes, L.M. 1996. Perception of speed in an automobile: Estimation and production. *Journal of Experimental Psychology*, 4(2): 291–304.

Rudin-Brown, C.M. 2004. Vehicle height affects drivers' speed perception: Implications for rollover risk. *Transportation Research Record*, 1899: 84–9.

Shinar, D. 1977. *Driver Visual Limitations, Diagnosis and Treatment* (Final Report on National Highway Traffic Safety Administration Contract No. DOT HS 5 1275). Washington, DC: US Department of Transportation.

Shinar, D. 2007. *Traffic Safety and Human Behavior*. Amsterdam: Elsevier.

Staplin, L., Lococo, K., Byington, S. and Harkey, D. 2001. *Highway Design Handbook for Older Drivers and Pedestrians* (Report FHWA-RD-01-103). Washington, DC: Federal Highway Administration.

Theeuwes, J., Alferdinck, J.W.A.M. and Perel, M. 2002. Relation between glare and driving performance. *Human Factors*, 44(1): 95–107.

Tijus, C., Barcenilla, J., Cambon de Lafayette, B. and Meunier, J.G. 2007. The design, understanding and usage of pictograms. *Studies in Writing*, 21: 17–32.

Walraven, J., Varkevissr, J. and Bijl, P. 1996. *Evaluatie van de leesbaarheid van een nieuw ANWB-alfabet voor bewegwijzering* [*Evaluation of the Legibility of a New ANWB Font for Guide Signs*] (TNO Report TM-96-C063). Soesterberg: TNO Technische Menskunde (in Dutch).

Wertheim, A.H. 1986. *Over het meten van visuele opvallendheid van objecten in het verkeer* [*On the Measuring of Visual Conspicuity of Objects in Traffic*] (TNO Report IZF 1986 C-25). Soesterberg: TNO Instituut voor Zintuigfysiologie (in Dutch).

Whelan, M., Langford, J., Oxley, J., Koppel, S. and Charlton, J. 2006. *The Elderly and Mobility: A Review of the Literature* (Report No. 255). Clayton, Victoria: Monash University Accident Research Centre.

Wilschut, E.S. 2009. *The Impact of In-Vehicle Information Systems on Simulated Driving Performance*. PhD thesis, University of Groningen, the Netherlands.

Chapter 7
Individual Information Elements are Consistent and Uniform within Their Context

Richard van der Horst

Individual information elements should be mutually consistent and uniform within their context, and should not confuse the driver.

Theoretical Background

Chapter 6 dealt with basic ergonomic principles of information carriers as such. But it is also important that information coming from different sources match each other to ensure that the same message is communicated to the road user. As indicated in Chapter 2, road user expectations – or 'expectancy' as Alexander and Lunenfeld (1986) called it – play an important role in conducting the traffic task. Alexander and Lunenfeld defined it as:

> Driver expectancy relates to the readiness of the driver to respond to events, situations, or the presentation of information. It is primarily a function of the driver's experience. When an expectancy is violated, longer response time and incorrect behaviour usually result.

If people encounter situations that meet their expectations (the situation is recognizable, they are locally well known, have experience of what they may encounter and know what is expected of them) they quickly know what to do, do not make errors and their behaviour is also predictable to other road users. In cases where people experience a situation different from what they expect, then the processing of information takes longer and they may display more hesitation, uncertainty and possibly incorrect behaviour.

Expectations have both a long-term and a short-term component (Alexander and Lunenfeld 1986). Long-term expectations build up during a lifelong experience and learning process. An example of short-term expectation is that recent experience of a certain type of intersection along a stretch of road also creates expectations for the next intersection to come.

As is extensively discussed in Chapter 2, people structure their world by gaining the maximum amount of information with as little cognitive effort as possible and develop internal prototypical representations. If the appearance of different road environments is homogeneous and maximally different from other types of roads, and the expected behaviour is explicitly coupled to that type of road (for example, motorways in the Netherlands: fast driving, no at-level intersections, no slow traffic etc.), then a consistent prototypical representation can easily develop. Good recognition is a necessary but not sufficient condition to induce safe behaviour, since road users should also be able and willing to display the behaviour intended by the roadway designer. For each road situation, it should be clear which speed limit applies, what type of intersection is to be expected, what route information is available and which road users one can expect and where. In short, for each road category, for all three hierarchical levels of the driving task, the traffic situation should meet road users' expectancies about the function and usage of that category of road. The expectancy road users develop relates to the design of the infrastructure (both roadway and traffic management) as well as to the intended behaviour of those users themselves and of other road users. For example, homogeneity in usage for a road with separate cycle lanes implies that one does not expect cyclists on the carriageway. Unexpected events simply require more time to be detected, perceived and interpreted, and for the road user to display the proper response/behaviour. This also implies that, in particular, transition zones from one road category to another deserve extra attention and more time for road users to adapt their behaviour (Wegman and Aarts 2006).

Research by Kaptein, van Hattum and van der Horst (1998) shows that people prefer to make use of only a few characteristics per category and a very limited number of categories – usually not more than two to four.

The Driving Context

Conflicting information or information other than expected may cause distraction, confusion, longer reaction times etc. For example, inconsistencies in destinations as displayed on successive route signs (one destination mentioned at the first sign and left out or another displayed at the next one) frequently confuse drivers and result in hesitation or incorrect behaviour by road users not familiar with the local situation.

Providing wrong or part-wrong information is sometimes worse than giving no information at all. Redundancy in providing information in traffic is important on the one hand to reach more people since different groups of road users base their judgement on different information elements (Kaptein et al. 1998; see also Chapter 3); and on the other hand in situations where information available can easily be overlooked, for example because of obstrution by a leading vehicle. The route signage strategy on motorways is a good example of the latter, with a pre-announcement of exit destinations at 1,200 metres before the exit, a repetition sign at 600 metres and information for both exiting and through traffic just before the decision point itself. People like confirmation that they have made the right decision. In cases where drivers remain uncertain about being on the right route, this may result in hesitant and errant behaviour after the decision point. They keep wondering whether or not they are on the right track and, consequently, this may negatively influence their performance at other levels of the driving task.

The dosing and the separating of information needed for conducting different subtasks may help prevent an overload of information in a too short a period of time.

Resulting Design Consequences

An important consequence of the distinction of the three hierarchical task levels in the driving task (see Chapter 4) – the information that each level requires and the notion that drivers are serial rather than parallel information processors (Lunenfeld and Alexander 1976) – is that individual information elements are presented unequivocally, unambiguously and conspicuously enough to enable drivers to make adequate decisions in traffic. Post, Alexander and Lunenfeld (1981) define the *decision sight distance* as the distance at which a driver can detect relevant information in an environment of visual noise or clutter, recognize it, select an appropriate speed and path, and perform the required action safely and efficiently. The main point here is that information carriers (location, timing and content) meet drivers expectancy. In this process recognition and comprehension are key issues. For recognition, the presented information has to fit in a set of elements the road user has already seen or experienced. Of course, the frequency may differ: some information elements are encountered daily or several times a day, and others only seldom. An information element/message might be conspicuous, easily recognizable and readable, but not understandable/comprehensible. The comprehensibility of a message can be expressed as the probability that a user relates it to the intended meaning and knows what to do with it.

Easy recognition of information relevant for route choice requires consistent and uniform application of route information dependent on the type of road one is driving on. Motorway driving at high speeds requires that drivers are being prepared for an upcoming route choice decision point; that the decision point is recognized as such; and that the information provided is timely readable and understood. An example of an information carrier whose appearance is not consistent with driver expectancy is a dynamic travel information panel installed along a motorway in the Netherlands a few years ago (see Figure 7.1). Such a panel should not have the general appearance of a route sign as it is only relevant for a specific group of long-distance travellers, and therefore may confuse others.

Also at the guidance level of the driving task, consistent and uniform application of specific characteristics in the context of a given road environment is important to ensure that road users behave safely and efficiently. For example, once one has made the decision to exit a motorway, the exit lane should be clearly visible, recognized as such and of sufficient length to deal properly with other traffic on the road. In the Netherlands motorways have a good, recognizable exit lane with block markings (see Figure 7.2), although sometimes there is confusion since entry lanes also have similar block markings and exit signs are already visible

Figure 7.1 Example of the misuse of a typical route sign appearance for a travel information sign park and ride information panel (PRIP)

Note: *Vertrek* = Departure; *Aankomst* = Arrival; Sittard, Utrecht and Den Haag are cities in the Netherlands; the logo in the right top corner represents the Dutch Railways.

at the entry area (Figure 7.3). From a driver's point of view it would be better to clearly make a distinction between both markings either by a different pattern or by adding exit or entry arrows on the road surface.

Specific delineation per road category (see Chapter 2) is another way to make the road category (and the behaviour expected) recognizable and comprehensible to the road user (see CROW 2004). But here also consistency and uniform

Figure 7.2 Example of specific road markings for exit lanes on Dutch motorways

Figure 7.3 Example of an entry lane with a route sign for the next exit already visible

Figure 7.4 Example of confusing road marking with both permanent and temporary markings in a work zone

application (also among different road authorities) is crucial from the road users' point of view. For example, if a green median is directly linked to a speed limit of 100 km/h, one should not design roads with a green median and a speed limit of 80 km/h or roads with a speed limit of 100 km/h without a green median.

Temporary delineation in work zones is implemented in yellow, but often the original road marking in white is still there (or visible). This may confuse drivers at the control level of the driving task, and it is recognized as one of the top ten irritation issues in traffic for road users (see Figure 7.4).

Figure 7.5 Example of combined information at both the navigation and guidance level of the driving task

Figure 7.6 Example of non-consistent or conflicting information to the road user from the traffic sign and the painted directional arrows on the road surface

Note: This is a still image taken from a video.

Consistency in design and application of uniform information provision may help drivers to better predict what is to be expected so they can anticipate a situation instead of being surprised at the wrong moment. This applies to information at all levels of the driving task, but also to information at one level that may help drivers prepare better for the next lower level. For example, arrows on route signs at the navigation level make the link with the guidance level by indicating schematically the layout of how the path for the current direction will continue (see Figure 7.5). The arrows on the route signs are identical to the arrows painted on the road surface and indicate directly the number of lanes available for each manoeuvre.

An example that does not meet the consistency principle is given in Figure 7.6, with the traffic sign indicating that only straight ahead is permitted (*dus niet afslaan* means 'so, do not turn'), whereas the arrows on the road surface show only left or right turning.

A somewhat new application of a consistent and uniform combination of information at both the navigation and guidance level is the redesign of route panels at interchanges on Dutch motorways. Instead of the pre-announcement of the current system (see Figure 7.7, top) that induces most road users to start using

Figure 7.7 Example of combined navigation and guidance information to improve traffic flow in the Netherlands

the right lane right away, it is already indicated at this point that the middle lane also provides access to the right turning carriageway (see Figure 7.7, bottom). This helps to optimize traffic flow. Initial tests revealed a 5 per cent increase in road capacity.

More and more, information is provided by dynamic devices – especially on the main network in the Netherlands – with all kinds of dynamic traffic management measures. On most motorways in the Netherlands an Automated Incident Detection (AID) System has been operational for a long time. This warns drivers of an oncoming traffic jam by flashing lights together with a speed sign of 70 km/h (as a kind of pre-warning) and of 50 km/h when the traffic has come to a standstill (see Figure 7.8).

These matrix signs are now used for other purposes as well, for example to indicate a lower speed limit in areas with high levels of emissions and poor air quality (see Figure 7.9). It is important to note that drivers also receive information when these signs are off: viz. no traffic jam ahead or no reduced speed limit. In the case of reduced speed limits on motorways that are designed for a speed of 120 km/h, all kinds of additional measures are needed to communicate messages to road users those that the roadway design itself induce. Traffic management measures and road design should be in line with each other; if not, a special effort has to be made to make it clear to drivers what behaviour is expected.

As also indicated in Chapter 3, uniformity and consistency are important to communicate with the road user to elicit the right behaviour unambiguously The examples given include the inconsistent meaning of the solid edge line when

Figure 7.8 Automated Incident Detection System on Dutch motorways

Figure 7.9 Example of reduced speed limits on Dutch motorways in air pollution sensitive areas

using the emergency lane as a rush-hour lane; two different appearances of speed limits on overhead matrix signs (with and without a red boundary circle); and the different communication of the 'stop' message at signalized intersections, movable bridges and railway level crossings (van der Horst 1988).

So-called 'Berm DRIPs' (roadside Dynamic Route Information Panels) are fully programmable electronic panels along the road that use a combination of symbols, pictograms and text (see Figure 7.10). With this type of provision it is a key issue to keep a good balance between what is technically possible and what can be properly handled by the road user. The configuration presented must coincide with the mental model the road user has from the network, and the level of detail presented requires a relatively long reading time and, consequently, huge signs along motorways. As it is implemented now, it does not meet the basic ergonomic principles as presented in the previous chapter. All the information that is given may confuse a driver who is not a frequent user of this network. For now, the information as presented may be only useful for drivers who regularly pass such a sign and have learned how to interpret the information given for their own individual situation.

The red-X is a recognized international symbol to indicate a lane closure and is understood well by 97 per cent of drivers according to Barten (2006). Unfortunately, in recent years, red-X non-compliance has risen, with an increased risk for people working on or at the side of a closed lane (emergency staff, road workers etc.). Increasingly, the red-X is also being used for lane control (for example rush-hour lanes) in situations where there is no physical blocking of the lane (see Figure 3.4). As reasons for the closure of a lane may differ, road users may form their own interpretation when relatively frequently confronted with the red-X for lane

Figure 7.10 Example of a 'Berm DRIP' (roadside Dynamic Route Information Panel)

Figure 7.11 Example of lane closure symbol for lane control in France

control, with a potential devaluation of the meaning of a red-X. From a human factors point of view it would be recommended to communicate these two reasons for lane closure through different symbols (Martens et al. 2010), for example with the symbol used in France (see Figure 7.11). This is an option worth examining in more detail for the Dutch motorway situation.

In-car navigation systems have become more and more popular and definitely fulfil users' needs. It is very important that the information the in-car system is providing matches the information shown along the road on traditional route signs with respect to distance, directional choice, destinations and lane choice. After a major reconstruction of the main road network around Eindhoven, for example, it appeared that the route maps of various in-car navigation systems were outdated and did not match the new situation, leading to a lot of confusion, erratic manoeuvres and long detours.

Summary and Conclusions

Information from different sources should match to ensure that the same message is communicated to the road user. Contradictory or conflicting information or information other than expected may cause distraction, confusion, longer reaction times etc. Information carriers (the location, the timing and the content) should meet driver expectations. Providing wrong or part-wrong information is sometimes worse than not giving any information at all. Consistency in the design and application of uniform information may help drivers to better predict what is expected and so anticipate what is coming instead of being surprised at the wrong moment. Several good and bad examples are provided to illustrate the relevance of tuning all information needed for the three hierarchical levels of the driving task in a consistent and uniform manner.

References

Alexander, G.J. and Lunenfeld, H. 1986. *Driver Expectancy in Highway Design and Traffic Operations* (Report FHWA-TO-86-1). Washington, DC: US Department of Transportation, Federal Highway Administration.

Barten, M. 2006. *Misbruik spitsstroken: Resultaten van een gebruikersonderzoek* (Report No. MV-SE20060789). Amersfoort: DHV (in Dutch).

CROW. 2004. *Richtlijn essentiële herkenbaarheidkenmerken van weginfrastructuur: Wegwijzer voor implementatie* (CROW publicatie 203). Ede: CROW.

Horst, A.R.A. van der. 1988. Driver Decision Making at Traffic Signals. *Transportation Research Record, 1172*. Washington, DC: National Academy Press, 93–7.

Kaptein, N.A., Hattum, S.T. van and Horst, A.R.A. van der 1998. *Categorization of Road Environments and Driving Speed* (MASTER Deliverable D9). Espoo, Finland: VTT Communities and Infrastructure.

Lunenfeld, H. and Alexander, G.J. 1976. *Improving the Highway Information System*. Washington, DC: Federal Highway Administration, Office of Traffic Operations.

Martens, M.H., Brouwer, R.F.T., Wilmink, I.R. and Horst, A.R.A. van der. 2010. *Korte en langere termijn oplossingen rood kruis negatie* (TNO Report TNO-DV 2010 C310). Soesterberg: TNO Human Factors Research Institute (in Dutch).

Post, T.J., Alexander, G.J. and Lunenfeld, H. 1981. *A Users' Guide to Positive Guidance*, 2nd edition (Report No. FHWA-TO-81-1). Washington, DC: Federal Highway Administration, Office of Traffic Operations.

Purcell, A.T. 1986. Environmental perception and affect: A schema discrepancy model. *Environment and Behavior*, 18: 3–30.

Wegman, F. and Aarts, L. (eds). 2006. *Advancing Sustainable Safety: National Road Safety Outlook for 2005–2020*. Leidschendam: SWOV Institute for Road Safety.

Chapter 8
Risk and Adverse Side Effects of Measures

Jan Theeuwes

When introducing particular measures it is likely that there will be adverse side effects. The driver is an integral part of the traffic system and will adapt their behaviour when necessary and possible. Any change within this system may lead to behaviour adaptation.

Theoretical Background

Over the last 25 years or so it has become increasingly clear that for each measure taken in road design there may be adverse side effects. The underlying notion is that people not simply 'undergo' changes in the environment, but they respond to changes in the environment by changing and adjusting their behaviour. From a safety point of view this implies that expected benefits of particular measures may not be as large as originally expected. This idea has been around for many years. In a paper written more than 70 years ago Gibson and Crooks (1938) argue that:

> More efficient brakes on an automobile will not in themselves make driving the automobile any safer. Better brakes will reduce the absolute size of the minimum stopping zone, it is true, but the driver soon learns this new zone and, since it is his field zone ratio which remains constant, he allows only the same relative margin between field and zone as before. (Gibson and Crooks 1938: 458)

The most compelling well-known evidence for compensatory mechanism comes from a study conducted in Germany involving taxicabs in Munich, half of which were equipped with anti-lock brakes (ABS) while the other half had conventional brakes. Even though from an engineering point of view the safety benefits of anti-lock brake are clear, it turned out that accident rates for both types of car (ABS and non-ABS) remained the same. It was argued that the ABS car drivers took more risks, assuming that the ABS would take care of them (Aschenbrenner and Biehl 1994).

The basic idea is that drivers will react to changes in the driving environment (in the car, on the road and through internal 'personality' settings and skills such as mood and capabilities) in a compensatory manner. This implies that they are willing to engage in riskier behaviour as a result of perceiving the driving environment as safer. From a traffic safety point of view, the key question is how much of this compensatory behaviour actually occurs. If it is complete, then not many safety benefits from a particular measure can be expected; if it is incomplete then a safety measure will have some effect. There are several theories that explain why safety benefits may be less effective than expected.

Risk Homeostatic Theory

The most influential theory to account for reduced safety benefits of particular measures is the risk homeostasis theory of Wilde (1994). According to this theory, in any activity people accept a certain level of estimated risk to the matters they value, such as health and safety, in exchange for the benefits they expect to receive from that activity. The idea is that people have the ability to check the amount of risk they are exposed to. They compare the estimated risk level to the amount of risk they are willing to accept and reduce the difference between these two levels to zero. If they feel that they are exposed to a relatively low level of risk, people will engage in actions that increase their exposure to risk. Alternatively, when they feel the experienced risk is too high, they change their behaviour such that the experienced risk becomes less and they typically behave with greater caution. In essence, the risk homeostasis theory claims the following:

> People alter their behaviour in response to the implementation of health and safety measures, but the riskiness of the way they behave will not change, unless those measures are capable of motivating people to alter the amount of risk they are willing to incur (=target level of risk). (Wilde 1994, online)

Crucially, according to this theory, unless the target level of risk changes, no permanent change in safety can be expected. Figure 8.1 presents Wilde's model. As is clear from this model, traffic safety only increases or decreases during periods in which the balance between perceived level risk and target level risk is not yet established (for example, because of lagged feedback mechanism). Wilde (1994) adheres the somewhat controversial viewpoint that the number of accidents in any given country depends only on the accident rate that the population is willing to tolerate and not on the specific measures taken, at least when considered over a longer period of time. The only way to gain safety benefits is to reduce the target level of risk (see Figure 8.1).

One of the famous examples of risk homeostasis comes from Sweden, where in 1967 traffic changed from driving on the left side of the road to the right. There was a major reduction in traffic fatalities and injury rates. Because people perceived a higher level of risk, they compensated by driving very carefully. After they became used to the new situation, accident rates returned to their usual levels (Trimpop 1996). The above-mentioned Munich taxicab example also fits perfectly with this theory. Drivers with ABS estimated the risk of getting involved in a collision as being lower than those without ABS and therefore adjusted their behaviour such that it became more risky. Because of the riskier behaviour there was no net safety effect of the ABS system. The conclusion is that in the end particular safety measures may have no effect because drivers adapt their behaviour in such

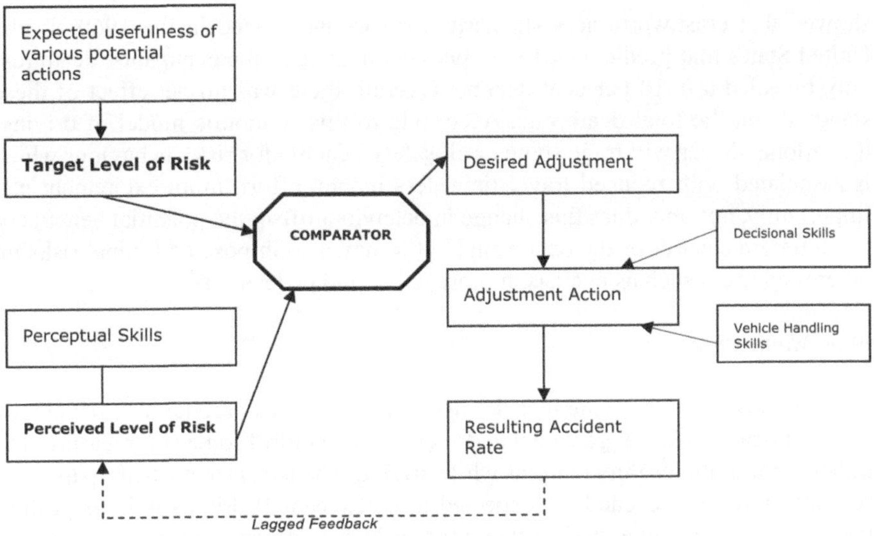

Figure 8.1 The risk homeostasis theory of Wilde (1994): Comparison between driver behaviour, accident rate and the level of target risk

a way that the potential benefits of a particular countermeasure are nullified by the behavioural adaptation.

To illustrate this with another example related to road design, in one village in the Netherlands, on a country road many trees were located very close to the main road. Each year at least three drivers were killed because they had hit one of these trees. One day the local authorities decided to cut down all the trees along this stretch of road to ensure that people could not be killed by driving into a tree. In the years following, still three drivers per year died but now not because of hitting trees but because of excess speeding on this stretch of road followed by loss of control of their vehicle. The underlying mechanism is clear. The presence of the trees was seen as a risk forcing people to slow down. When the trees were cut down there was no longer any risk and drivers adjusted their behaviour and started speeding. In the end the net safety effect was zero. There are numerous similar anecdotal examples and empirical studies showing that behavioural adaptation occurs after particular countermeasures are taken (see Wilde 2001). In addition to accident statistics showing risk homeostasis, experimental evidence also suggests that drivers will compensate.

Risk Compensation

The theory of risk compensation (for example, Peltzman 1975) is basically an economic approach to determine the effects of traffic safety measures. This theory is very similar to the theory of risk homeostasis. In his classic study, Peltzman showed that crashworthiness standards for cars introduced in the 1960s in the United States that predicted a 15–35 per cent decline in the occupants' death rate only revealed a 0–10 per cent decline. Overall, there was no net effect of these standards on the total death rate. According to this economic model of driving, the rational driver will trade some or all safety benefits for risky behaviour which is associated with reduced travel time, less mental effort, more excitement etc. Importantly, not only does this change in behaviour offset the potential benefits of the countermeasure for the occupants, but it may also impose additional risks on other road users such as cyclists, motorcyclists and pedestrians.

Fear Monitoring

Another version of the same idea, known as the 'zero-risk' model, was developed by Summala (1988), suggesting that drivers do not control some risk measures but rather monitor their safety margins while driving. The idea is that when a particular fear threshold is exceeded (or expected to be exceeded) drivers will adapt their behaviour. It is the hypothetical 'fear monitor' which alarms and influences driver decisions when safety margin thresholds are violated. This model deviates from the models above in that it assume that drivers may perceive particular situations initially as 'fearful', but after prolonged exposure to them they no longer consider particular safety margins as risky and drive most of the time on the basis of 'overlearned' behavioural patterns. Also, this model incorporates the idea that particular motives which may exist outside the immediate momentary traffic situation may affect willingness to take risks. For example, when someone is late for an appointment other thresholds are used to evaluate risk than when there is no hurry at all. It is important to realize that particular external circumstances (such as being rushed, tired or frustrated) may affect the level of risk accepted, and therefore will affect the extent to which behavioural adaptation occurs.

Some Empirical Evidence for Behavioural Adaptation

This section briefly reviews data that indicate that risk compensation will take place when a particular measure is introduced.

Mandatory Seat Belt Legislation

It is well known that when a crash occurs, wearing a seat belt will reduce quite substantially the probability of death and/or injury. On average the probability of a vehicle's occupants being killed or seriously injured falls by about by 40–50 per cent (Elvik and Vaa 2004). Even though these figures are quite dramatic, there has been no substantial drop in accident rates after legislation has been implemented in any given country. These disappointing results have invariably been attributed to some kind of behavioural compensation (for example, Evans

1991; Hurst 1979; Janssen 1994). Because drivers feel safer when strapped into a seat belt, they drive less carefully. The idea that risk compensation is at play is confirmed by studies that showed that car occupant fatalities increased after the adoption of mandatory seat belt laws (Hurst 1979). Because, in the past, typically the non-driving occupants did not wear seat belts, the driver driving less carefully (because they were wearing a seat belt) would selectively influence the chances of the non-seat belt wearing occupants getting hurt or killed. Moreover, this study also showed that the implementation of the legislation resulted in a substantial increase in fatalities of other road users – again evidence of risk compensation.

One of the very few experimental investigations into behavioural adaptation was conducted by Janssen (1994). He investigated the on-road behaviour of drivers who did and not habitually wear seat belts on a motorway in the Netherlands. The results showed increased speeds among those drivers who normally did not wear seat belts and were forced to use a seat belt during the experiment. In the follow-up study, it was shown that first-time seat belt wearers showed signs of continuing behavioural adaptation, as evidenced by increased speeds and an increased tendency to follow too closely behind other vehicles (tailgating).

Anti-lock Braking System (ABS)

As discussed earlier, the Aschenbrenner and Biehl (1994) study on Munich taxi drivers showed that the accident rate of taxis equipped with ABS was no lower – and, if anything, was higher – than those of taxis not equipped with ABS. It was reported that drivers changed their behaviour quite dramatically: drivers of taxis equipped with ABS made sharper turns on curves, swerved out of their lanes more often and had shorter following distances to the car in front.

An experimental study conducted by Grant and Smiley (1993) confirms the notion that behavioural adaptation may play a role. In this experiment, test drivers were exposed to both ABS and non-ABS conditions involving normal and emergency braking. The results showed that test drivers with ABS drove faster, hit the brakes harder and accelerated faster. It was concluded that ABS may result is less safe driving.

Airbags

There is also some evidence that airbags installed in a vehicle will result in behavioural adaptation. For example, based on insurance data and accident reports, one study shows that drivers of cars with airbags drive more aggressively than when no airbag is installed (Peterson, Hoffer and Millner 1995). Even though this appears to be some evidence of behavioural adaptation, it should be realized that airbags are less 'visible' than, for example, an ABS system – of which drivers are much more aware. For example, each time a driver hits the brakes in slippery weather the ABS system kicks in. If airbags do result in compensatory behaviour this can only work at a strategic level. For example, drivers with airbag-equipped cars may be less likely to wear a seat belt when they start to drive. In a study by Sagberg, Fosser and Saetermo (1997) the effects of airbags were compared to ABS and the results indicated that ABS, but not airbags, are related to behavioural adaptation in terms of tailgating. This also fits with the notion that measures that are targeted at reducing injury severity will result in less behavioural adaptation than measures that directly affect driving safety (see below).

Considerations

Following publications in the early 1980s, there has been quite some controversy about risk homeostasis and compensation. Some have claimed that the theory is incoherent in its theoretical formulations (Haight 1986); others even claimed that 'the tone of advocacy for the claim has been largely philosophical, metaphysical, and theological in nature, unencumbered by the standards, methods, or norms of science' (Evans 1991: 299). Others have criticized the idea of risk as a central variable (for example, McKenna 1988; Rothengatter 1988), while yet others indicated that the theory does not allow one to generate testable hypotheses (for example, Hoyes and Glendon 1993; Ranney 1994).

In 1990 an international working group presented a report on behalf of the Organisation for Economic Co-operation and Development (OECD) on risk compensation. They concluded that:

> behavioural adaptation to road safety programmes does occur although not consistently ... behavioural adaptation generally does not eliminate the safety gains from programmes, but tends to reduce the size of the expected effects.

So even though it is clear that risk compensation occurs, there is little support for the extreme view that there is always a full adaptation nullifying the effect of particular road safety measures.

Even though the extreme view of risk compensation may not be convincing, it is clear that the driver is an integral part of the traffic system and any change within the system will result in more or less behaviour adaptation. It is important to realize that it is impossible to estimate the potential benefits of one measure without considering adverse side effects. For example, research shows that public lighting of motorways in the Netherlands increased the comfort of driving and reduced workload. However, in rainy conditions drivers chose a much higher speed when the motorways were lit than when they were unlit (Hogema and Kaptein 1998), providing evidence that risk compensation reduced the potential positive effects of public lighting. Similar findings were reported by Assum et al. (1999), who showed that road lighting resulted in increased speed, made road users less alert and that road users who formerly did not travel on unlit roads started driving once the roads were lit. It was concluded that this study can be considered evidence for behavioural adaptation partly (or possibly fully) offsetting the engineering effect of placing lights along the road.

More empirically oriented researchers argue that the risk homeostasis theory is too vague to test empirically. Elvik (2004) argues for a set of more specific hypotheses regarding risk adaptation that one should be able to test empirically. He proposes six factors (adapted from Bjørnskau and Amundsen 2003) that are assumed to affect behavioural adaptation. Also, in the OECD report various measures are mentioned which are less likely to cause behavioural compensation. Below we discuss some of these considerations.

How Easily a Measure is Noticed (Visibility of a Measure)

The underlying idea is that road users constantly scan the road environment and will notice those aspects of the environment that has changed (Elvik 2004). For instance, the above-mentioned example of cutting down the trees along a road is a highly 'visible' measure and is likely to result in a change of behaviour, such as an increase in speed. Similarly, increasing the sight distance by removing obstacles (taking out curves) is also a measure that is likely to result in behavioural adaption because drivers will experience an increased safety margin. On the other hand, if a car manufacturer introduced a collapsible steering column, drivers may not know this and are not likely to adapt their behaviour. Also, if a car manufacturer were to introduce a clever safety measure such as skid-, traction-control and roll-over protection systems without telling the customer, the effectiveness would be much higher. Obviously, a car manufacturer likes to sell its cars by promoting how safe they are, while in fact not telling customers about these features would make the cars safer. In this respect it is interesting to consider the effect of advertisements to promote safe behaviour. For example, pointing out through advertisements – including information presented along the motorway (so-called public service information signs) – that using a seat belt has many safety benefits will ultimately reduce their effectiveness. In this respect it would be better to show that wearing a seat belt is required by law without referring to its safety benefits. The bottom line is that changes to the road environment or the vehicle that are not easily noticed are much more effective because there is a smaller chance that behavioural adaptation will take place.

Antecedent Behavioural Adaptation to Basic Risk Factors

This idea suggested by Elvik (2004) addresses the concern that when drivers already took particular risks into account (displaying antecedent behavioural adaptation) and these risks are reduced, then it is likely that drivers will adapt their behaviour. For example, if I drive very carefully in wintry conditions because I have not changed my vehicle's regular tyres to snow tyres, I am highly likely to change my behaviour dramatically as soon as the snow tyres are put on. Behavioural compensation will be large. However, if there is more road lighting at night, behavioural compensation will be less dramatic because drivers do not perceive driving at night as a risk factor, at least not a risk factor that makes them drive very slowly on unlit, dark roads. In other words, if one takes a measure that is assumed to affect a risk that drivers already take into account, then large behavioural adaption is expected.

The Greater the Changes Made, the Larger Behavioural Adaptation is Expected

This is also a factor mentioned by Elvik (2004) and is in fact related to the point mentioned above which dealt with the ease with which a particular measure is noticed. For example, when a separate cycle lane is added to a route, drivers will immediately notice that cyclists are no longer riding on the main road. This will typically lead to large behavioural adaptation – for example, drivers will drive faster because there is physical separation between them and the cyclists. More subtle measures such as improving road design by making the roads more self-explaining (SER) – as discussed in Chapter 2 – will make the roads safer without much, if any, behavioural adaptation (Theeuwes and Godthelp 1995). In a good SER design, the road looks the way it should look and this should induce a completely different behavioural repertoire, one that is not affected by risk adaptation.

Measures Reducing Injury Severity Will Result in Less Behavioural Adaptation

Measures that focus purely on reducing the consequences of a collision are less likely to affect behavioural adaption (Elvik 2004) because, typically, drivers try to circumvent collisions altogether. For example, as discussed earlier, a driver is less likely to start speeding because he has an airbag (which can only affect the impact of a collision).

The Smaller the Effect on Impact, the Smaller the Behavioural Adaptation

The notion here is that when there is less damage (to the vehicle or person) on impact the less road users will adapt. A driver in a heavy sports utility vehicle (SUV) surrounded by tons of steel is less likely to adapt (drive safer, wear a seat belt or attend to the road) than a cyclist who may be killed on impact.

Behavioural Adaptation as a Result of Utility Maximisation

The risk compensation theory assumes that drivers constantly adapt their behaviour such that the perceived risk stays more or less the same. Although feasible, it should be realized that drivers seek out behavioural optimization, and being safe is just one of the aims for the driver. There may be other goals, such as getting to a meeting on time, having lunch in the car while driving or speaking on the phone. Because these goals compete with the goal of driving safely, drivers may adapt their behaviour to maximize utility.

The Continuity of the Perception of Risk

If a particular measure does not affect the perception of risk then it is unlikely that behavioural compensation will occur. Risk perception during driving is a rather complex process and for behavioural adaptation to occur one has to assume that the estimation of risk is a process that constantly takes place. Even though it is clear that for some measures such continuous estimation is likely to occur (for example, a seat belt and or safety helmet is constantly present), for other measures which are not constantly present it seems unlikely that risk is constantly estimated. For example, drivers may not be aware that the car they are driving has skid-control and anti-lock brakes.

Automatic Behaviour Pattern Will Be Less Sensitive to Behavioural Adaption

Theories of behavioural adaptation (and risk monitoring) implicitly assume that human behaviour is rational and assume a major role for cognitive processes. Clearly this is certainly not the case for driving, as most tasks are conducted in an automatic way. This implies that drivers like to simplify decisions and often act on the basis of automatic behavioural patterns. When behaviour is more or less automatic (such as looking over the shoulder when turning right) it is unlikely that this will go into the risk compensation equation.

Enforcement

Enforcement is one of the most effective ways to overcome behavioural compensation. For example, if the presence of lighting along a motorway is compensated for by an increase in driving speed, the simplest way to counteract risk compensation is to enforce a speed limit. There is a substantial body of literature showing that enforcement is an effective way to ensure adequate behaviour. For example, improving road conditions by making lanes wider, putting up barriers, making line markings more visible by improving contrast, installing public lighting etc will definitely result in higher driving speeds. In this case the only way to ensure adequate driving speeds is to implement speed enforcement. The literature is clear that the most important aspect of enforcement is to increase the perceived risk of apprehension. In general, the cost-effectiveness of enforcement programmes related to excess speed, drinking and driving and seat belt use is invariably positive (Zaal 1994).

Wilde (1994) suggests four approaches to reduce the target level risk for drivers (see Figure 8.1):

1. Drivers may be rewarded for particular safe behaviour.
2. Drivers may be rewarded for not crashing during a particular period of time (as insurance companies do with insurance premiums).

3. Punishing unsafe behaviour. Fines for speeding in urban areas (where children may be playing and cyclists may be crossing) may be more effective because speeding in urban areas is seen as much more dangerous than speeding on the motorway. In this way, unsafe behaviour may be punished. In this respect, putting penalty points on driving licences for unsafe behaviour may also work as a punishment.
4. Punishing drivers who have crashed (also implicitly through insurance premiums).

Increase the Perception of Risk

Even though few studies have addressed this aspect of the risk compensation theory, one way to ensure safer driver behaviour is to increase the perception of risk without raising the actual, objective risk. To give an example: instead of having actual trees along the roadway, one could place faux-concrete blocks along the road that appear dangerous while in fact nothing would happen if someone hit them. In a study conducted in the Netherlands similar ideas were applied on some country roads in the north of the country (De Waard et al. 1995). Typically, drivers tend to speed on these types of road. Some of the measures taken included 'visually' reducing the lane width so that the lane looked narrower while in fact the actual road surface remained the same. Moreover, instead of visible lane markings on the right-hand side of the road, there were wide unpainted and intermittent edge lines made of gravel. This gravel would not give visual guidance but would alarm drivers when exiting their lane. This study showed that, relative to a control condition with conventional road design, drivers reduced their speed significantly. This study illustrates that making a road appear more risky may in fact have safety benefits. Whether such effects are long-lasting is not clear because it is possible that over time drivers may realize that the risk is not as great as these roads subjectively suggest. In any event, instead of trying to introduce safety measures that are basically invisible, it is an interesting option to introduce highly visible measures that appear to increase the risk while in fact they do not.

Other Benefits

Even though most measures are directed towards increasing traffic safety, it should be realized that due to risk compensation particular measures may have little effect on traffic safety but may have other potentially positive effects such as increased traffic capacity or increased comfort of driving (for example, reduced workload). For instance, using public lighting on Dutch roads may have had little effect on safety but research has shown that public lighting reduced driver workload considerably (Hogema and Veltman 2002).

Side Effects of Measures Not Related to Risk Compensation

The analysis above focused on the side effects of safety measures by means of the mechanism of risk compensation. It should be realized, however, that there may be other side effects of particular measures which may be unrelated to risk compensation. Recall that in the Netherlands, due to high congestion, during rush hours all traffic can use the emergency lane as a regular lane. This in itself is suboptimal because it induces some level of uncertainty on the part of the driver on how to behave during rush hour (Can I move into the lane? Can I overtake while in this lane? At what speed should I drive? etc.). But, more importantly, this measure may induce adverse side effects. Because this measure allows drivers to cross the solid marker line on the right-hand side of the road to get into this lane, drivers learn that it is acceptable to cross solid road markings. This implies that overlearned behaviour repertoires such as 'never cross a white solid line' will lose their strength. The consequence is that in other places, in other circumstances, drivers may be more likely to cross a solid line because they have learned that it is okay to do so. Another example is the 'red cross' on variable message signs (VMS) above the road to indicate that a lane cannot be used. In the past this was only used when a lane was blocked due to a collision. Nowadays, however, in the Netherlands the red cross is used for other reasons, such as indicating that particular lanes cannot be used when traffic density is relatively low. In other words the red cross, which in the past unequivocally meant 'accident and/or danger', has lost its strength because it is now used for every occasion in which a lane cannot be used. Again this illustrates that particular measures can have adverse side effects, affecting the behaviour of drivers in subtle ways.

Summary and Conclusions

When introducing particular measures, it is important to look at the whole road system and consider which side effects are likely to occur. It is clear that risk compensation does occur but, dependent on the measure (for example, its salience and what it impacts), such compensation is absolute or only partial. For some measures no compensation is expected. Note that due to risk compensation and/or due to drivers violating particular overlearned behavioural patterns adverse side effects may occur in such a way that there are safety costs. In other cases there may be no safety benefits but there may be other benefits, such as improved comfort or increased traffic throughput. It is important to balance the pros and cons whenever a measure is introduced. It is important to never assume that there will be no behavioural changes associated with introducing new measures.

References

Aschenbrenner, M. and Biehl, B. 1994. Improved safety through improved technical measures? Empirical studies regarding risk compensation processes in relation to anti-lock braking systems. In: R.M. Trimpop and G.J.S. Wilde (eds), *Challenges to Accident Prevention: The Issue of Risk Compensation Behaviour*. Groningen: Styx Publications, 81–90.

Assum, T., Bjornskau, T., Fosser, S. and Sagberg, F. 1999. Risk compensation: The case of road lighting. *Accident Analysis and Prevention*, 31(5): 545–53.

De Waard, D., Jessurun, M., Steyvers, F.J.J.M., Raggatt, P.T.F. and Brookhuis, K.A. 1995. Effect of road layout and road environment on driving performance, drivers' physiology and road appreciation. *Ergonomics*, 38: 1395–407.

Elvik, R. 2004. To what extent can theory account for the findings of road safety evaluation studies? *Accident Analysis and Prevention*, 36(5): 841–9.

Elvik, R. and Vaa, T. 2004. *The Handbook of Road Safety Measures*. Oxford: Elsevier.

Evans, L. 1991 *Traffic Safety and the Driver*. New York: Van Nostrand Reinhold.

Evans, L. and Gerrish, P.H. 1996. Antilock brakes and risk of front and rear impact in two-vehicle crashes. *Accident Analysis and Prevention*, 28(3): 315–23.

Fuller, R. 1986. Reflections on risk homeostasis theory. In: B. Brehmer, H. Jungermann, P. Lourens and G. Seven (eds), *New Directions in Research on Decision Making*. Amsterdam: North-Holland, 263–73.

Gibson, J.J. and Crooks, L.E. 1938. A theoretical field-analysis of automobile driving. *American Journal of Psychology*, 51: 453–71.

Grant, B.A. and Smiley, A. 1993. *Driver Response to Antilock Brakes: A Demonstration on Behavioral Adaptation.* Proceedings, Canadian Multidisciplinary Road Safety Conference VIII, Saskatchewan, 14–16 June.

Haight, F.A. 1986. Risk, especially risk of a traffic accident. *Accident Analysis and Prevention*, 18(5): 359–66.

Hogema, J.H. and Kaptein, N.A. 1998. *Dynamic Road-Lighting (DYNO). Phase 3: Evaluation in Practice* (Report TM98C038). Soesterberg: TNO Human Factors Research Institute.

Hogema, J.H. and Veltman, J.A. 2002. *Werkbelasting en rijgedrag tijdens duisternis: eerste veldexperiment* (Report TM-02-C046). Soesterberg: TNO Technische Menskunde.

Hoyes, T.W. and Glendon, A.I. 1993. Risk homeostasis: Issues for future research. *Safety Science*, 16: 19–33.

Hurst, P. 1979. Compulsory seat belt use: Further inferences. *Accident Analysis and Prevention*, 11(1): 27–33.

Janssen, W. 1994. Seat belt wearing and driving behaviour: An instrumented-vehicle study. *Accident Analysis and Prevention*, 26(2): 249–61.

McKenna, F.P. 1988. What role should the concept of risk play in theories of accident involvement? *Ergonomics*, 31: 469–84.

Organisation for Economic Co-operation and Development (OECD). 1990. *Behavioural Adaptations to Changes in the Road Transport System*. Paris: OECD.

Peltzman, S. 1975. The effects of automobile safety regulation. *Journal of Political Economy*, 83: 677–725.

Peterson, S., Hoffer, G. and Millner, E. 1995. Are drivers of air-bag-equipped cars more aggressive? A test of the offsetting behavior hypothesis. *Journal of Law and Economics*, 38: 251–64.

Ranney, T.A. 1994. Models of driving behavior: A review of their evolution. *Accident Analysis and Prevention*, 26(6): 733–50.

Rothengatter, T. 1988. Risk and the absence of pleasure: A motivational approach to modelling road user behaviour. *Ergonomics*, 31: 599–607.

Sagberg, F., Fosser, S. and Saetermo, I.F. 1997. An investigation of behavioural adaptation to airbags and antilock brakes among taxi drivers. *Accident Analysis and Prevention*, 29(3): 293–302.

Summala, H. 1988. Risk control is not risk adjustment: The zero-risk theory of driver behaviour and its implications. *Ergonomics*, 31: 491–506.

Trimpop, R.M. 1996. Risk homoeostasis theory: Problems of the past and promises for the future. *Safety Science*, 22: 119–30.

Van der Horst, A.R.A. and Hoekstra, W. 1994. Testing speed reduction designs for 80 kilometre per hour roads with simulator. *Transportation Research Record*, 1464: 63–8.

Wilde, G.J.S. 2001. *Target Risk 2*. Toronto: PDE Publications.

Wilde, G.J.S. 1994. *Target Risk: Dealing with the Danger of Death, Disease and Damage in Everyday Decisions*. Toronto: PDE Publications. Available at: http://psyc.queensu.ca/target/chapter01.html [accessed: 18 April 2012].

Wilde, G.J.S. 1982. The theory of risk homeostasis: Implications for safety and health. *Risk Analysis*, 2: 209–58.

Zaal, D. 1994. *Traffic Law Enforcement: A Review of the Literature*. Clayton, Victoria: Monash University Accident Research Centre.

Chapter 9
Motivational Aspects

Maria Kuiken

The road user maximizes their behaviour to reach goals and adapts their behaviour towards that end. Non-driving activities such as telephone conversations, entertainment, relaxation and work are also included in this process.

Theoretical Background

Aberrant Behaviour

A core attribute of a well-designed traffic system is the compatibility between the formal regulations and actual behaviour on the road. Reality shows that this compatibility can only be partially realized: road users have a tendency to deviate from traffic norms and rules, and from the optimal level of task performance. This behaviour is either unintended (an error or lapse) or it is a conscious act (a violation). By making a distinction between error and violation a distinction is made between performance approaches (experimental psychology) and motivational approaches (social psychology). The distinction is relevant for behavioural scientists because underlying behavioural mechanisms and also remedies differ.

Unintended aberrant behaviour such as driver error is mainly concerned with human factors and characteristics of the road system. In driving, as much as in other activities, people will usually try to attain goals effectively, efficiently and safely. In other words, they will try to behave intelligently (Michon, Smiley and Aasman 1990). From this perspective, driving is similar to problem solving – that is to reach a goal by means of successively diminishing the distance between a present state and a goal state in a problem space that allows certain actions but not others (Newell and Simon 1972). Much of the behaviour displayed by drivers can be regarded as adequate responses to momentary road and traffic situations. But errors occur. Some theories attribute error tendencies to innate or acquired traits (see also Chapter 10). Others address the functionality and the lack of functionality of system performance that may induce erroneous behaviour and collisions (see most of the previous chapters in this book). Drivers must cope with a considerable amount of information. This information concerns all three levels of the driving task: vehicle control at the operational level; manoeuvring at the tactical level; and navigation at the strategic level (Michon 1985). The road and traffic environment may evoke errors on all three levels, especially in drivers who are prone to error, because they are, for instance, inexperienced, distracted or tired.

Social Cognition

Notwithstanding the frequent occurrence of errors, road users also deviate deliberately from traffic norms and rules. They violate traffic rules and regulations consistently and make reckless manoeuvres. This behaviour – which applies equally to drivers, cyclists and pedestrians – poses disproportional risk. Speeding, tailgating, running red lights, illegal use of restricted lanes, failure to yield and illegal manoeuvres such as U-turns are the most frequently cited offences, and a large proportion of collisions appear to be preceded by at least one of these offences. Why do drivers violate traffic rules and norms? What motivates a person to speed or to use a restricted lane?

From a social psychological perspective social cognitions are important. These include constructs such as motives, attitude, attribution and cognitive biases. These constructs reflect clearly defined individual (psychological) characteristics that are latent and cannot directly be observed (Steg and Rothengatter 2008). Attitude generally refers to the thoughts and feelings that impel us to behave in one way and not in another. Social psychological theories hold that the best way to affect long-lasting change in behaviour is to change the beliefs, values and attitudes that underpin the decision to behave in a certain way (Parker 2004). The Theory of Planned Behaviour (TPB) is particularly relevant in this context, as it has also been applied to road user behaviour. According to motivational models a driver can make the driving task easier, for example by reducing speed or increasing headway (De Craen 2010). Examples of motivational models that primarily focus on the self-pacing aspects of the driving task are: Wilde's (1982) risk homeostasis theory, the zero risk theory of Summala and Naatanen (1988), the risk avoidance model (Fuller 1984), the task-capability model (Fuller 2005) and the task-difficulty homeostasis model (Fuller 2008).

Attitudes and Social Norms

Compliance with traffic regulations has a positive effect on traffic safety (Elvik 2004). Elvik found that compliance with speed limits and proper use of protective equipment generates the largest impact on traffic safety. Large-scale research on differences among European driver attitudes is the Social Attitudes to Road Traffic Risk in Europe (SARTRE) project. The researchers investigated, among other things, the degree of strictness in matters of traffic safety. Table 9.1 shows the extent of the differences between various questions and countries with respect to this dimension.

The results show a north–south line. The southern European countries tend to emphasize personal freedom; northern countries seem more ready to accept regulations with respect to drink/driving, lights in the daytime and speed. Vrolix and Vereeck (2006) investigated differences in road safety levels in the European Union (EU) and also focused on the importance of attitudes of European citizens towards road safety policy and legislation. The authors investigated four

Table 9.1 **Percentage of respondents agreeing with different measures (from SARTRE 1995: 41)**

Country	Obligation to use lights daytime	Decide self drink/drive	If careful belts not necessary	Maximum speed limit on motorways in km/h	
	In favour	*Strongly agree*	*Agree*	*110*	*140–50*
Sweden	91%	2%	4%	47%	2%
Denmark	86%	2%	14%	17%	4%
France	14%	30%	21%	6%	31%
Italy	13%	27%	31%	7%	30%
Portugal	23%	27%	27%	4%	23%

explanatory variables that might explain the variance in road fatalities between European countries:

- The Corruption Perceptions Index (CPI), which is a proxy for the general attitudes and social norms of citizens towards traffic legislation and policy. This index gives an indication of the tolerance of the population towards evasion of rules or in general the moral climate in a country. The CPI had a negative effect on the number of road fatalities.
- Alcohol consumption per capita, which was used as a proxy for the drinking habits of a population. This variable had a positive effect on traffic fatalities.
- Gross domestic product (GDP) per capita, the economic factor, was used as an indication of road infrastructure and vehicle conditions. The better the economic situation, the better traffic safety.
- Road traffic volume, an exposure variable, shows a positive effect on the number of traffic fatalities.

The authors concluded that northern European countries seem to be more successful in improving and maintaining their road safety standards.

Theory of Planned Behaviour (TPB)

The Theory of Planned Behaviour (Ajzen 1985, 1988) proposes that behaviour results from intentions (see Figure 9.1). Intentions are strong when people have a favourable attitude towards the behaviour (that is, the degree to which engagement of certain behaviour is positively valued); when they think others expect them to engage in that behaviour (for example, subjective norm); and when they think they are capable of performing the relevant behaviour (for example, perceived behaviour control).

The TPB has been applied to virtually every conceivable type of road user behaviour and has been successful in predicting these behaviours (for example Parker, Lajunen and Stradling 1998; Stradling and Parker 1997; Rothengatter 1994; Yagil 1998). Parker (2004) concludes that the TPB has proved a useful way

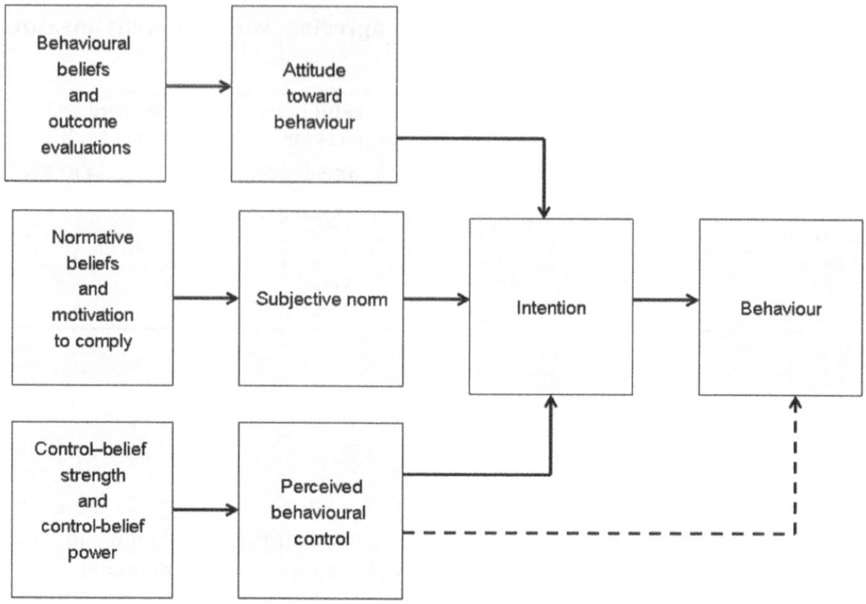

Figure 9.1 Theory of Planned Behaviour (after Ajzen 1988)

of identifying beliefs and attitudes typical of the individual who behaves in a risky fashion.

However, there was also some criticism. Rothengatter (2002) indicated for instance that the application of attitude theories seems to occur out of the environmental context. The theory presumes that the components in the model (attitude, social norm and perceived behavioural control) determine the intention to perform a specific behaviour at the exclusion of other factors, including environmental variables. This does not hold. The design of our motorways is clearly one of the factors that influence driver behaviour. Engineering principles sometimes conflict with human perceptions and behaviour, and we must be aware of these influences and attempt to offset or accommodate these in designing safer roads (Fuller 2002). Studying the relationship between environmental factors, the task environment and behaviour can link attitude theory to other realms of traffic psychology and this may provide a useful and useable framework to study the elusive mechanisms of behavioural adaptation. The relationship between task environment and attitude model components and their impact on behaviour needs further study.

Another criticism of TPB is that our understanding of the motives of road users that influence deviant and dangerous behaviour has contributed very little to measures to remedy this behaviour. Very few (successful) interventions have been developed thus far that target these motivations. Can attitudes be changed? Årberg reports that many attempts have been made to influence road user attitudes by means of mass-media campaigns. Elliott (1993) analysed 87 campaigns that were properly evaluated. Findings show that although campaigns had effects on behaviour, the effects on attitude were moderate. There are examples that show a relationship between attitudes and behaviour, but Årberg concludes that this relationship might be quite complex. The predictive power of general attitudes towards traffic safety can be expected to be quite low.

Other social constructs that have been studied in relation to road user behaviour involve cognitive biases, for example attribution and biases in the perception of self. The theory holds that cognitive biases systematically colour our thinking. Attribution refers to the way in which we understand and interpret the behaviour of others. This bias is particularly relevant with respect to aggressive behaviour. Other biases in perception have been related to self-enhancement, optimism bias, illusion of control and self-justification. Individuals see themselves as more skilful than others and they think they have more control over their own behaviour and the environment than is actually the case.

Cognitive biases have been investigated in the context of driving. A well-known attribution error is that drivers tend to attribute their own behaviour to situational factors, such as being in a hurry, and ascribe the behaviour of the other driver to dispositional factors, such a being an aggressive person. Parker (2004) gives an overview of studies that demonstrate this tendency. In summarizing the results Parker indicates that one could argue that attempts to inform and educate the general public about these cognitive biases could be valuable: for example sources of frustration and aggression can be remedied by altering drivers' subjective perceptions of the situation they find themselves in. On the other hand, several researchers have shown that countering these biases is not easy.

Adaptive Behaviour and External Motives or Tasks or Motives

Motivational models assume that driving is a self-paced task and that drivers determine the amount of risk or difficulty they are willing to take. The theories indicate that road users are able to weigh a risk against other possible benefits of their behaviour – such as the amount of effort they will have to use, time profit, pleasure of driving (fast), comfort or any other factor that may move them. By the same token improvements in the road environment can be used by road users to adapt to the changed environment and optimize their individual benefit.

External Motive or Task Can Decrease Adaptivity

Drivers have to monitor the environment continuously to avoid collisions with other vehicles and stationary objects. Experimental studies (for example, Van der Hulst, Meijman, and Rothengatter 1999, 2001) support the hypothesis that drivers use safety margins that determine for instance their time headway to a lead vehicle. Safety margins reflect the amount of time drivers allow themselves in the interaction with other road users and the environment. Adaptation of speed and safety margins to current traffic situations allows drivers to control time pressure. The time margins chosen in actual driving are possibly more relevant for safe driving than the driver's actual competence (Summala 1996).

The detection of relevant information on which the safety margins are based depends not only on perceptual factors but also on the expectations of the driver. Based on experience, drivers know when relevant information is likely to appear and what other road users are likely to do in the near future. Van der Hulst et al. (1999) showed in a simulator study that drivers' expectations affect the minimum time headway in several deceleration scenarios. When a lead car decelerated unexpectedly, headway was much shorter than when it decelerated expectedly. The findings suggest that drivers try to avoid the high time pressure associated with short headways. Their actions were directed at maintaining the preferred time margin to control the time available to react to potential hazards. The authors conclude that several factors can hinder this anticipatory driving strategy. Sometimes environmental cues are not clear and reliable (for example due to fog and darkness); sometimes distraction can affect the possibility to build expectations.

The study supports the hypothesis that, at the operational level of task performance, external motives or tasks such as a decreased visibility, time pressure, talking to a passenger or telephoning could reduce the adaptation of the driver. The motive or tasks may temporarily overrule other goals in driving and can induce a decreased flexibility of driving strategy. In these situations driver behaviour generally becomes less adaptive to prevailing circumstances. This is particularly relevant with respect to the use of a mobile phone while driving. The majority of drivers report using their mobile phone while driving. Telephoning is an additional task that causes a physical and cognitive distraction. The driver has to divert part of their attention from driving to the telephone conversation. The distractive effects of mobile phone use depend on the momentary context of driving. There may only be a problem in situations with a high task demand. This may lead to slower reactions to traffic signals, or no reaction at all, slower braking reactions and reduced general awareness of other traffic (Dragutinovic and Twisk 2005).

External Motive or Task and Compensatory Behaviour

By the same token, external motives or tasks can also lead to compensatory behaviour. In a simulator study by Cnossen, Rothengatter and Meijman (2000) subjects had to drive in various speed conditions with and without a memory task. The results show that, even in difficult driving situations, drivers can still achieve a high level of task performance when performing a number of tasks concurrently. Task behaviour appeared to be strategic. Drivers focused more on relevant information – for example, navigation information – than on an artificial memory task. Tasks are prioritized with respect to the main task goal.

Van der Hulst et al. (2001) focused on changes in performance in driving in relation to feelings of fatigue and aversion. These can also be considered as 'external tasks'. A positive relation was found between levels of fatigue and time headway: fatigued drivers increased their time headway more than drivers who did not report an increase in fatigue. Drivers clearly try to avoid the high costs associated with a particular (external) task, either by skipping the less important subtasks or by adapting their driving behaviour. They will attempt to prioritize their task goals and, if necessary, make the driving task easier. The studies demonstrate that drivers are perfectly capable of dropping the additional external tasks if the driving task with immediate safety relevance requires attention. This is consistent with Brown's (1994) statement that prolonged driving will not necessarily cause serious problems if drivers can choose the speed and the safety margins they prefer and if they can stop driving if they want to. The main concern should therefore be with where these adaptive responses are not possible, either because of high traffic densities, high speeds or because of strict time schedules of the professional driver (Cnossen et al. 2000).

Fuller's Task-Capability Model: Optimizing Risk, Effort and Utility

It is not always easy to distinguish between intended and unintended behaviour; they are often interrelated events. To give an example: disregarding traffic lights is always a violation in legal terms and not an error, even if the driver did not intend to disregard the red light. When one fails to notice a red traffic light, it is regarded as an excuse which will not help in avoiding being fined. Also other behaviours – for example, failure to yield – may be attributed to performance or violation. Moreover, the consequence of the one is aggravated by the other. It can be argued, for example, that speeding per se does not affect collision risk. However, it is very difficult to maintain that the consequences of error are the same at high speed as they are at low speed. Hence, there is a good argument for considering error and violation as complementary rather than opposing concepts, which interrelation should be subjected to study (Rothengatter 1997a).

Fuller (2000, 2002, 2005) proposed a task-capability model (TCM) in which drivers match their capability and performance with task demands to maintain control. This model focuses on the adaptive mechanisms in driving. One of the basic notions is that crashing is more or less continuously inevitable unless a driver does something about it. Fuller proposed already in an earlier model that drivers adjust their speed to deal *more easily* with some hazards or potential difficulty. The TCM goes a step further. The model is not situation specific. Driving is considered a dynamic control task in which the driver has to select relevant information from a vast array of (mainly) visual inputs to make decisions and execute appropriate control responses in order to achieve mobility with safety (Fuller 2005). The main model components are: the Task demand, the Driver (capability/performance) and the Task-capability interface (see Figure 9.2). Task difficulty arises out of the dynamic interface between the demands of the driving task and the capability of

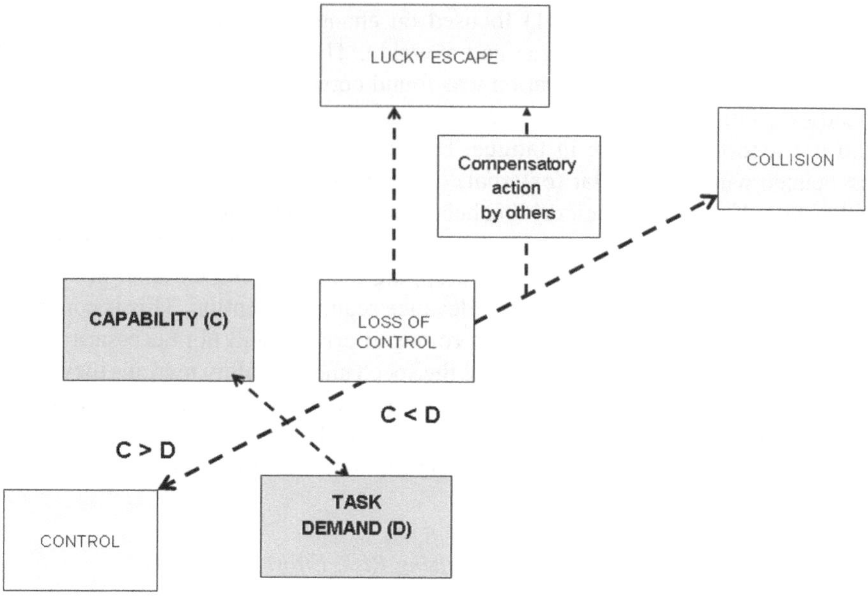

Figure 9.2 The Task-Capability Interface Model: Outcomes of the dynamic interface between task demand and capability (after Fuller 2005)

the driver. If the capability of a driver exceeds the task demand, the driver is able to progress safely. If capability is less than the task demands, then loss of control is implied.

Fuller (2002: 7) hypothesizes that:

> for most of the time drivers drive so as to achieve their mobility, their travel goals, while ensuring the *difficulty* of the task remains within acceptable limits. Thus if things are getting too hectic on the road, too demanding, the driver slows down. If the task is boringly easy, the driver speeds up, making it more challenging.

An important characteristic of capability is the ability to predict correctly potential hazards on the road. Individuals differ in this ability 'to read the road'. Inexperienced drivers are more often in a 'reactive' mode of dealing with hazards; experienced drivers are more likely to deal with hazards in an anticipatory way. This model seems more productive than the earlier conceptualizations in terms of risk considerations because it considers drivers' actions in performance terms instead of risk terms (Rothengatter 2002). Risk is an illusive concept. It is much easier to establish what parameters drivers utilize to assess the competence–demand match.

Drivers do not always match competence and task demand such that they have optimal control. In discussing a later version of the task-capability interface (TCI) model, Fuller indicates that the capability is vulnerable to human factors variables – such as factors of attitude, motivation, effort, fatigue, distraction, emotion and stress – and that any of these can lead to a lower level of capability or a compromise (Fuller 2005). Drivers can influence task demands by making choices in relation to each of the factors. Performing additional tasks can push demand beyond driver capability. A simple task can become challenging to a degree where adaptive behaviour is necessary.

In short, Fuller's TCI model takes account of the environment and specific driving situations; and it takes account of driver capabilities, including motivational aspects. It states that drivers are motivated to maintain a preferred level of task difficulty, and use compensatory behaviours to realize their goals. Fuller's model incorporates the hierarchical nature of the driving task. Drivers can make the driving task easier on different levels. They can take a strategic decision (for example, choosing another route) or they can increase margins at the control level (for example, decreasing speed or increasing headway). The driver might also decide to 'increase' their own capabilities by changing external conditions (for example, music or temperature) or internal conditions (for example, alertness and focus). A crucial aspect is however that the models assume that the driver is capable of making the correct assessment of the task demands and his or her capabilities. Otherwise they will not see the need to adapt their behaviour (De Craen 2010). Recently, Fuller (2008) has adapted his model into the task-difficulty homeostasis model. This new model incorporates the assessment part and focuses especially on adaptation at the lower levels of task performance: drivers' choice of speed and trajectory. The balancing of task demands and skills has been referred to as 'calibration' (Fuller 2008; Kuiken and Twisk 2001; De Craen 2010).

Implications from a Social Perspective

Road Users Are Adaptive Intelligent Problem Solvers

Road users are intelligent problem solvers, or at least they have the capability to be so. Their behaviour is in principle flexible and adaptive. The adaptation is influenced by driving related motives and tasks, and by external motives and tasks. Road users have different capabilities and they are adaptive in different ways. They can adapt their behaviour at different levels: at strategic level or at lower levels of task performance. Road users are adaptive in their destination/goal, in their travel pattern, in their route choice; people are able to neatly distribute their resources between operational and tactical behaviours.

It needs to be considered that there may be situations where the collective intelligence of road users may be more effective in solving the problems on the road than using centralized traffic regulation. One of the problems is of course that they optimize their own goals, not the collective goals. The use of their collective intelligence will not solve safety problems on the road. However, it may lead to a more adaptive system that is tolerant of incidents, errors and breakdown. People are able to optimize their situation within explicit constraints.

Engineers become increasingly aware that their expectations about the effectiveness of (traffic management or safety) measures need to be moderated. When a traffic situation changes, drivers generally adapt their behaviour. Sometimes this behaviour is unexpected and counter-effective. An interesting example of such behaviour occurred on Dutch motorways. The speed limits on motorways were, in some areas around cities, reduced from 120 to 80 km/h. An enforcement system was set in place by using trajectory control. Drivers received a fine when their mean speed over a particular trajectory was higher than the speed limit. The result was that, after some time – and particularly in areas with a high traffic density – drivers drove considerably slower than the 80 km/h they were allowed to drive at. This had a negative impact on throughput and caused a loss of 'dynamics' in traffic, which made weaving and lane change more difficult. This negative side effect was probably due to the fact that drivers normally do not calculate their mean speed. It is an additional task that takes effort. It is not easy to monitor your speed over a long period. Consequently drivers tend to choose a somewhat lower speed and give up part of their 'active' driving behaviour. They reduce their speed and number of manoeuvres, and settle for the fact that they have to drive much slower and most likely begin to perform other (non-driving) tasks.

Adaptivity depends on information: the gathering of information, the quality of the information and the limits in processing the information. For road users to be adaptive and to be able to use their problem solving capability, they need to obtain the right kind of information. Information management, both at a road system level and on an individual level, is one of the core challenges in our road and traffic system. Adaptation puts great pressure on the need to integrate information from road systems, the vehicle and personalized agents.

Making use of the intelligence of road users themselves requires the integration of information from the road and road systems, in-car systems and personalized agents such as global positioning systems (GPS) and telephones. Roads and road systems must become 'knowledge factories' and provide the information so that the driver remains 'in the loop', even in complex situations.

The Role of Reward and Punishment

Motivation plays an important role in traffic behaviour and is a strong influence on whether people will comply with regulations. If compliance with traffic regulations is not close to 100 per cent, the regulations must be enforced. It is recommended to combine two strategies for enforcement (Siegrist and Roskova 2001).

The *first* strategy concerns the use of a minimal level of highly visible enforcement. Results of studies on the road indicate that various forms of observable police surveillance can have an effect on traffic behaviour. This effect is most pronounced when there is an actual presence of police surveillance. However experiments carried out in the 1980s demonstrated that intensifying apparent enforcement without increasing the objective level of apprehension does not influence actual speed choice, or the motivations and attitudes towards speed choice (Rothengatter 1988). In situations where the objective level of apprehension was increased by adding radar check posts to the surveillance activity, a marked reduction in speed violations was achieved.

The *second* strategy concerns the detection of non-compliance. Below a certain level, enforcement is inconsequential; increases or decreases will not affect driver behaviour (Rothengatter 1997b). If enforcement can be carried out automatically the subjective risk of detection can be elevated to a level at which it influences behaviour. Automatic enforcement has proved to be technically viable and very effective in reducing collision occurrence; but the acceptance of such systems is a major issue, with very contradictory aspects. On the one hand, drivers resist the idea that their freedom is to be curbed; on the other hand, drivers applaud when other drivers' reckless behaviour is detected and punished (Rothengatter and Harper 1991).

An issue that demands closer attention is that enforcement does affect behaviour but, at least initially, does not affect attitudes towards that behaviour. This has become apparent in studies concerning drunk-driving. Only when enforcement persists over a period of years, and enforcement is coupled with massive media campaigns, do attitudes adapt to the enforced behaviour. This mechanism, which is extremely important to effective traffic law enforcement, is not yet understood (Rothengatter 1997b).

Also rewards can be powerful external motivators for behaviour (Rothengatter 1992; Hagenzieker 1999). After a series of experiments Hagenzieker concluded that: reward programmes are especially effective in small-scale settings; the programmes are especially effective in reaching substantial short-term effects; and immediate delivery of rewards is generally more effective than delayed rewards (see also Chapter 11).

Implications for Designing Safe Roads

The Task-Capability Interface model of Fuller has an immediate relevance for the motorway engineer. Fuller describes three implications of the model.

The *first* implication is that motorway engineers should try to avoid that drivers perceived task difficulty differs from the objective task difficulty. Underestimation of the task difficulty can easily lead to a situation where task demands exceed the capabilities of the driver. There are two strategies to prevent this from occurring. One is that the driver should learn not to underestimate the situation by training and experience. The other is that drivers are presented with clear and reliable information – information about the appropriate speed, alignment, an immediate hazard ahead etc. It is particularly important to facilitate a driver's anticipation of oncoming situations and events that could occur.

The *second* implication is that motorway engineers should try to make the task appear more difficult than it objectively is. A driver may, for instance, choose to increase their speed to shorten the journey without getting the sense that they actually thereby increase the task difficulty. Road design should prevent this from occurring.

The *third* implication of Fuller's model is that when a driver meets conditions where the task demands exceed their capabilities and can do nothing to reduce those demands or enhance their capability, the driver will attempt to avoid these conditions (Fuller 2002: 8). This explains why certain drivers may avoid particular routes or times of day, or opt out of driving altogether (see also Chapter 10).

Recommendations for motorway engineers from a social psychological perspective are:

- Know where and how drivers tend to deviate from a reference performance and be proactive: make the system fault tolerant.
- Provide road information (warnings, pre-warnings) from which drivers can anticipate (on strategic, tactical and control levels). Make it clear when there is an exception to a rule. Frequent behaviour is very strong and drivers need time and warnings to appreciate that the situation has changed.
- Be specific and explicit about the constraints of the road systems and the operational limits, particularly where safety is concerned, and do not allow drivers to operate outside these constraints; use enforcement.
- Invest in the development of a self-organizing system and information management. The finding that drivers are likely to increase their safety margins when they perceive the need for this (for example, when they become fatigued or have an additional task) has practical implications for the design of information systems in road traffic. The provision of high-quality performance monitoring and performance information is beneficial for safe driving. When a monitoring system detects performance impairments, it can warn the driver. The main concern should be with where these adaptive responses are not possible, either because of high traffic densities and/or high speeds or because of strict time schedules of the professional driver.

Summary and Conclusions

From a social psychological perspective, we have argued that drivers have their own goals and that drivers will maximize their behaviour. It can be assumed that the avoidance of collisions has the highest priority in driving. Drivers match capability and task demand and try to maintain in all situations an acceptable task difficulty. Drivers can restrict the task demands by relaxing their performance criteria in vehicle control and by increasing their safety margins if the task constraints allow them to do so. However, in regulating road user behaviour, road authorities and engineers must be aware that people will fill in their 'free space'.

In-vehicle technologies – such as telephones, navigation systems and infotainment systems – are increasingly common.

It is recommended that motorway engineers take account of this, for instance by making the system fault tolerant and by being specific and explicit about the constraints of the road systems. Adaptation puts great pressure on the need to integrate information from road systems, the vehicle and personalized agents. Making use of the intelligence of road users themselves requires the integration of information from the road and road systems, in-car systems and personalized agents such as GPS and telephones. Roads and road systems must be become 'knowledge factories' and provide information so that the driver remains 'in the loop', even in complex situations.

However, the strategy changes of drivers also depend on their motivation and are strongly influenced by the energy or effort a person is willing to use in the driving task. For safety reasons it is important that drivers invest in the primary task (driving). Attempts to inform and educate the general public about the cognitive biases they possess could be valuable by altering drivers' subjective perceptions of the situation they find themselves in.

References

Åberg, L. 1998. Traffic rules and traffic safety. *Safety Sciences*, 29(3): 205–15.

Ajzen, I. 1988. *Attitudes, Personality, and Behavior*. Chicago, IL: Dorsey Press.

Ajzen, I. 1985. From intentions to actions: A theory of planned behavior. In: J. Kuhl and J. Beckmann (eds), *Action Control: From Cognition to Behavior*. Berlin: Springer-Verlag, 11–39.

Brown, I.D. 1994. Driver fatigue. *Human Factors*, 36: 298–314.

Cnossen, F., Rothengatter, J.A. and Meijman, T.F. 2000. Strategic changes in task performance in simulated car driving as an adaptive response to task demands. *Transportation Research F*, 3: 123–40.

De Craen, S. 2010. *The X-Factor*. Leidschendam: SWOV Institute for Road Safety Research.

Dragutinovic, N. and Twisk, D. 2005. *Use of Mobile Phones while Driving: Effects on Road Safety* (R-2005-12). Leidschendam: SWOV Institute for Road Safety Research.

Elliott, B. 1993. *Road Safety Mass Media Campaigns: A Meta Analysis* (CR 118). Canberra: Federal Office of Road Safety.

Elvik, R. 2004. To what extent can theory account for the findings of road safety evaluation studies? *Accident Analysis and Prevention*, 36(5): 841–9.

Elvik, R. and Vaa, T. 2004. *The Handbook of Road Safety Measures*. Oxford: Elsevier.

Fuller, R. 1984. A conceptualization of driving behaviour as threat avoidance. *Ergonomics*, 27: 1139–55.

Fuller, R. 2000. The task-capability interface model of the driving process. *Recherche, Transports, Security*, 66: 47–59.

Fuller, R. 2002. Psychology and the highway engineer. In: R. Fuller and J.A. Santos (eds), *Human Factors for Highway Engineers*. Oxford: Elsevier.

Fuller, R. 2005. Towards a general theory of driver behaviour. *Accident Analysis and Prevention*, 37(3): 461–72.

Fuller, R. 2008. Driver training and assessment: Implications of the task difficulty homeostasis model. In: L. Dorn (ed), *Driver Behaviour and Training*, vol. III. Aldershot: Ashgate, 337–48.

Fuller, R. 2011. Driver control theory: From task difficulty homeostasis to risk allostasis. In: B. Porter (ed.), *Handbook of Traffic Psychology*. Amsterdam: Elsevier, 13–26.

Hagenzieker, M. 1999. Rewards and road user behaviour: An investigation of the effects of reward programs on safety belt use. PhD thesis, Leiden, Rijksuniversiteit.

Hulst, M. van der, Meijman, T.F. and Rothengatter, T. 1999. Anticipation and the adaptive control of safety margins in driving. *Ergonomics*, 42: 336–45.

Hulst, M. van der, Meijman, T.F. and Rothengatter, T. 2001. Maintaining task set under fatigue: A study on time-of-task effects in simulated driving. *Transportation Research F*, 4: 103–18.

Kuiken, M.J. and Twisk, D.A.M. 2001. *Safe Driving and the Training of Calibration: A Literature Review* (SWOV Report R-2001-29). Leidschendam: SWOV Institute for Road Safety Research.

Michon, J.A. 1985. A critical view of driver behavior models: What do we know, what should we do? In: L. Evans and R. Schwing (eds), *Human Behavior and Traffic Safety*. New York: Plenum Press.

Michon, J.A., Smiley, A. and Aasman, J. 1990. Errors and driver support systems. *Ergonomics*, 33(10–11): 1215–29.

Newell, A. and Simon, H.A. 1972. *Human Problem Solving*. Englewood Cliffs, NJ: Prentice Hall.

Parker, D. 2004. Road safety: What has social psychology to offer? In: J.A. Rothengatter and R.D. Huguenin (eds), *Traffic and Transport Psychology*. Oxford: Elsevier, 125–34.

Parker, D., Lajunen, T. and Stradling, S. 1998. Attitudinal predictors of interpersonally aggressive violations on the road. *Transportation Research F*, 1: 11–24.

Rothengatter, J.A. 1988. Risk and the absence of pleasure: A motivational approach to modeling road user behaviour. *Ergonomics*, 31(4): 599–607.

Rothengatter, J.A. 1992. The effects of police surveillance and enforcement on driver behaviour. *Current Psychological Reviews*, 2: 349–58.

Rothengatter, J.A. 1994. *Attitudes Towards Law Violation and Enforcement*. Proceedings of the 37th American Association of Automotive Medicine (AAAM) Conference. Portland, OR: AAAM.

Rothengatter, J.A. 1997a. Psychological aspects of road user behaviour. *Applied Psychology: An International Review*, 46(3): 223–34.

Rothengatter, J.A. 1997b. Errors and violations as factors in accident causation. In: J.A. Rothengatter and E. Carbonell Vaya (eds), *Traffic and Transport Psychology: Theory and Application*. Oxford: Elsevier.

Rothengatter, J.A. 2002. Drivers' illusions – no more risk. *Transportation Research F*, 5: 249–58.

Rothengatter, J.A. and Harper, J. 1991. The scope and design of automatic policing and information systems with limited artificial intelligence. In: *Advanced Telematics in Road Transport*. Proceedings of DRIVE Conference, Brussels. Amsterdam: Elsevier, 1499–515.

Siegrist, S. and Roskova, E. 2001. The effects of safety regulations and law enforcement. In: P.E. Barjonet, *Traffic Psychology Today*. London: Kluwer Academic Publishers.

SARTRE 1. 1995. *Social Attitudes to Road Traffic Risk in Europe: European Drivers and Traffic Safety – In-Depth Analyses*. Orléans: Paradigme.

Steg, L. and Rothengatter, T. 2008. Introduction to applied social psychology. In: L. Steg, A.P. Buunk and T. Rothengatter, *Applied Social Psychology: Understanding and Managing Social Problems*. Cambridge: Cambridge University Press, 1–27.

Stradling, S. and Parker, D. 1997. Extending the theory of planned behaviour: The role of personal norm, instrumental beliefs and affective beliefs in predicting driving violations. In: T. Rothengatter and E. Carbonell Vaya (eds), *Traffic and transport psychology: Theory and Application*. Oxford: Elsevier, 367–74.

Summala, H. 1996. Accident risk and driver behaviour. *Safety Science*, 22: 103–17.

Summala, H. and Naatanen, R. 1988. The zero-risk theory and overtaking decisions. In: T. Rothengatter and R. de Bruin (eds), *Road User Behaviour: Theory and Research*. Assen: Van Gorcum, 82–92.

Yagil, D. 1998. Gender and age-related differences in attitudes toward traffic law and traffic violations. *Transportation Research F*, 1: 123–35.

Vrolix, K. and Vereeck, L. 2006. *Social Norms and Traffic Safety: A Cross Country Analysis in the EU-15*. Steunpunt Beleidsrelevant Onderzoek. Verkeersveiligheid.

Wegman, F. and Aarts, L. 2006. *Advancing Sustainable Safety: National Road Safety Outlook for 2005–2020*. Leidschendam: SWOV Institute for Road Safety Research.

Wilde, G.J.S. 1982. The theory of risk homeostasis: Implications for safety and health. *Risk Analysis*, 2: 209–58.

Chapter 10
Variation in Performance

Maria Kuiken

Road users have different capabilities, are influenced by social factors and their performance varies over time. In road design and traffic management differences between groups should be taken into account.

Theoretical Background

The design of the road system is based on a prototype driver with some assumed set of characteristics and capabilities. It also assumes that driver behaviour is more or less constant over time and place. It is clear that in practice these assumptions do not hold. There are considerable differences between drivers, and human performance and motivation vary over time. Some of the variance in performance is caused by aspects of the environment. A continuous monotonous environment has a different effect on the level of activation of a driver than an environment with many stimuli and a high task demand. Road users not adapt only to environmental factors but also to individual factors. Their performance is vulnerable to a wide range of influences encompassing individual differences and external factors (Fuller 2001). Individual differences are based on constitutional characteristics (for example, characteristics of the driver vary as a function of age) and on differences in knowledge and skills (for example, characteristics of the driver vary as a function of experience). The external factors that play a role encompass task-generated factors (for example, emotions in traffic and factors related to time on task such as fatigue) and non-task generated factor (for example, level of aggression, use of drugs and alcohol).

These factors influence driving style and cognitions as well as motivations regarding driving. In this chapter the following themes and groups are discussed:

- Older drivers: age-related limitations.
- Young inexperienced drivers.
- Emotions and aggressive driving behaviour.

The Older Road User: Age-Related Limitations

The distribution of age groups is changing (see Table 10.1). The total population of EU-25 will fall slightly, but will become much older. The percentage of 65+ will be around 29 per cent of the total population.

Table 10.1 Projections of main European demographic trends (from Eurostat 2004)

Projections for the EU's population trend 2005–50				
in thousands	2005–50	2005–10	2010–30	2030–50
Total population	-8,659	5,563	5,312	-19,534
Percentage change	*-1.9%*	*1.2%*	*1.1%*	*4.2%*
Children (0–14)	-13,811	-2,304	-6,080	-5,427
Percentage change	*-18.6%*	*-3.1%*	*-8.5%*	*-8.2%*
Young people (15–24)	-14,035	-2,383	-6,663	-4,990
Percentage change	*-24.3%*	*-4.1%*	*-12.0%*	*-10.2%*
Young adults (25–39)	-24,867	-3,896	-14,883	-6,088
Percentage change	*-25.0%*	*-3.9%*	*-15.6%*	*-7.5%*
Adults (40–54)	-18,666	4,116	-10,029	-12,754
Percentage change	*-19.0%*	*4.1%*	*-9.8%*	*-13.8%*
Older workers (55–64)	4,721	4,973	8,717	-8,969
Percentage change	*9.1%*	*9.5%*	*15.3%*	*-13.6%*
Elderly people (65–79)	25,688	1,947	2,2281	1,460
Percentage change	*44.5%*	*3.4%*	*37.3%*	*1.8%*
Frail elderly (80+)	32,311	3,109	11,969	17,233
Percentage change	*171.6%*	*16.5%*	*54.0%*	*50.8%*

It is expected that the participation of older people in traffic will increase strongly over the coming years. The kilometres travelled by older drivers have already increased substantially during the past 20 years. Men and women over 70 years old will increasingly possess a driving licence.

Research has shown that perceptual, motor and cognitive changes cause deterioration in the task performance of older road users. In general, the main feature of ageing is the progressive slowness of behaviour (Simões and Marin-Lamellet 2002). However it should be realized that age-related behavioural changes occur at different levels and rhythms, leading to high variability among people who are elderly.

Functional Limitations of the Older Road User

Perceptual Abilities

The visual system becomes impaired. The eyesight worsens; the older person becomes less sensitive to visual contrast and is more impaired by visual glare (for example, oncoming headlights). The eye needs more time to accommodate to changing light conditions and the effects of glare also take longer to disappear again. Glare can also prevent the reading of signals. It becomes more difficult to drive at night (in darkness); older drivers need more light. Because of a diminished contrast-sensitivity, it will take more effort (and time) to read traffic signs, see depth and estimate the speed of other road users. Older drivers need more time for decision making and, generally, they will become slower (Wickens et al. 2004).

Cognitive Abilities

There are a number of cognitive changes that lead to functional limitations. Davidse (2007) describes a number of these cognitive functions, including: speed of processing, fluid intelligence, working memory and executive functions such as inhibition, flexibility and selective and divided attention. Older drivers have more difficulty suppressing information that is not relevant for the task. Consequently they have more difficulty in searching the visual environment for relevant information. This means that they will be more prone to failure in detecting target information – for instance, another road user. However, research studies have generally found that reaction times to simple stimuli do not deteriorate dramatically with age (Olson and Sivak 1986). Reaction time of older drivers only slows down when drivers have to make decisions in complex situations (Quimby and Watts 1981).

Motor Skills

Older traffic participants who have serious problems with their motor skills can be confronted with serious mobility problems. In 2001–02 one in ten people between 65 and 74 years suffered such problems with motor skills. Less severe problems concern reduction in joint flexibility, muscular strength and manual dexterity. These changes in motor skills – for example, reduced neck rotation – can influence their task performance (Davidse 2007).

Difficulties of Older Drivers

From accident data it is clear that the following crash types are overrepresented in the accidents that older drivers are considered legally responsible for (Davidse 2007):

- crashes at intersections (often because they fail to yield and especially when the driver has to turn left;
- collisions caused by fatigue and/or illness;
- joining/exiting through-traffic.

Research studies indicate that the most important driving difficulties and weaknesses that older drivers experience are the following (based on a review of Davidse 2007):

- difficulty judging whether fellow road users are moving and at what speed they approach intersections (motion perception and contrast sensitivity);
- overlooking other road users while merging and changing lanes (peripheral vision and flexibility of head and neck);
- overlooking traffic signs and signals (selective attention);
- reaction time increases as the complexity of the traffic situation increases (speed of processing information and decision making, divided attention, performance under pressure of time).

Compensation Strategies

These deficits do not only lead to an overrepresentation in collisions, they can also impose limitations on the ability of people to participate in traffic. The influence of age-related functional differences can be reduced in several ways. One of them is that the older traffic participants compensate for their own functional limitations. According to Fuller's task-capability interface model (2002a), people cope with discrepancies between the demands of the task and their capabilities. They make sure that the difficulty of the task remains within acceptable limits. Thus if the task is getting too complex, the driver slows down.

Older road users are experienced and this will help them to adjust their behaviour on the road. They make use of compensation mechanisms in order to adequately perform in traffic. This will allow them to suffer fewer limitations in their mobility. Possibilities for compensatory behaviour are mostly available on the higher task levels.

- At the strategic level a road user can decide when and where to travel. Within limits, strategic traffic decisions allow people to compensate for lower-level perceptual-motor impairment (for example by avoiding rush hours, avoiding driving at night) (Brouwer, Rothengatter and van Wolffelaar 1988).
- At the tactical level a driver can decide to keep more distance from a lead car; they leave bigger gaps and can decide to drive slower (see also Chapter 9).

Also, their experience allows them to be proactive, avoiding complex traffic situations (Simões and Marin-Lamellet 2002). In general, older people compensate with a progressive slowness of behaviour.

Table 10.2 Functional limitations and relevant design principles for road design elements

Functional limitations	Relevant design principles
Visual acuity	Increase character size of textual labels on traffic signs.
Night-time visual acuity	Use retro-reflective road signs,and fixed lighting at intersections with high traffic or pedestrian volumes.
Sensitivity to glare	Reduce the intensity of traffic signals during darkness.
Hearing	Use auditory signals in the 1,500–2,500 Hz range.
Contrast sensitivity and motion perception	Use a minimum in-service contrast level of 3.0 for pavement markings; background plates to help accentuate traffic lights; and a positive offset of opposing left-turn lanes.
Selective attention	Enhance the conspicuousness of critical stimuli through changes of size, contrast, colour or motion.
Perception–reaction time	Give the user sufficient time to respond to a request through using multiple or advance traffic signs.

Note: Based on Davidse 2007 (from Caird et al. 1998; Gardner-Bonneau and Gosbee 1997; and Staplin et al. 2001).

Implications for Road Design

What helps the older driver is a well-structured road and traffic system. It needs to be clear what is expected of a driver and where the relevant information can be found. Simões and Marin-Lamellet (2002) argue that this is also why it is so important to signal potential hazards such as a sharp bend after a long stretch of road.

The influence of age-related functional differences can also be reduced when the traffic task is made easier by simplifying traffic situations. There are measures that reduce task demands on the driver and increase decision time, such as replacing a crossroad with two T-junctions or replacing a crossroad with a mini roundabout. Another useful measure is to improve sight lines by removing visual obstruction.

Furthermore, Simões and Marin-Lamellet (2002) stress the importance of advance information with adequate size, lightning and glare protection for legibility for a number of concrete traffic situations (for example, driving onto or off motorways, roadworks. They also indicate that with regard to roadworks, drivers who are elderly require advance and clear roadside information.

Davidse (2007) provides an overview of functional limitations and relevant design principles for road design elements and in-vehicle displays. Table 10.2 shows only those limitations that are related to road design elements.

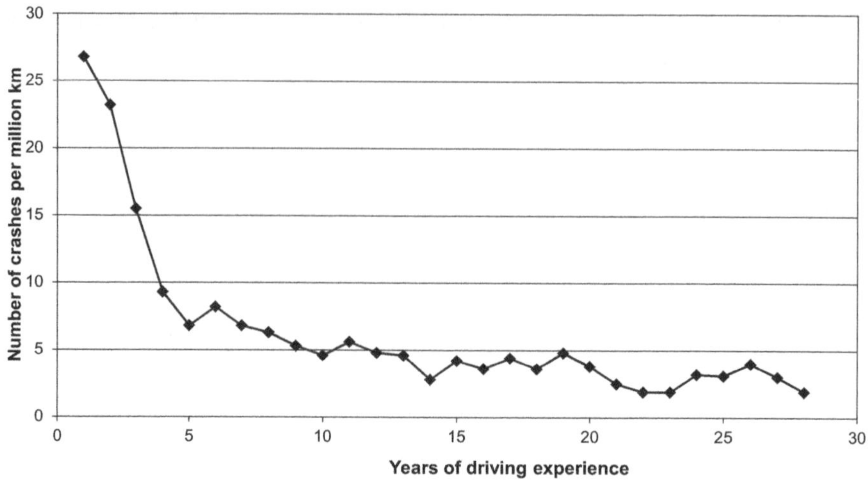

Figure 10.1 Crash risk and years (from Wegman and Aarts 2006)

Young and Inexperienced Drivers

Characteristics of Inexperienced Drivers

The effects of age and experience are powerful factors in collision involvement. Young drivers have more and more serious crashes per kilometre driven. The difference between experienced and inexperienced drivers can be explained in roughly two ways: capabilities and motivation (see Figure 10.1).

Capabilities

Inexperienced drivers still have difficulty in performing certain tasks.

- There are *deficiencies in competence*: they have less developed control skills and make more errors. Inexperienced drivers have a limited automation of driving subtasks (Fuller 2002a; Groeger 2000, 2006). For many driving situations this may lead to a higher mental workload for inexperienced drivers compared to experienced drivers (De Waard 2002; Patten et al. 2006). Generally, a distinction is made between two modes of information processing: automatic and controlled (Shiffrin and Schneider 1977). In the automatic actions are triggered automatically without the necessity for control or attention. The process is fast and fairly effortless. Controlled processing is slow and effortful. It is mostly used to deal with new or inconsistent information. Large subtasks of driving are regarded as 'automatic'. With practice more and more correct actions can be triggered with automatic processing, and driving a car then becomes less effortful. Overall, experienced drivers utilize their expectation of the traffic environment to anticipate what is about to happen next. This provides them with more time and space to respond to the situation, and reduces error. Knowledge about traffic grows with experience. The more routine a task becomes, the less mental workload is required to perform it. Inexperienced drivers are therefore more likely to produce the 'wrong solution' when faced with a novel situation (De Craen 2010).

- Inexperienced drivers have a strong tendency to *overestimate their own skills* and their ability to correct a situation. Clarke, Ward and Truman (2002) investigated a sample of 3,437 accident cases involving drivers aged 17–25 years covering the years 1994–96. They found that single-vehicle accidents, and in particular loss of control on bends, were a problem for the younger drivers. The proportion of single-vehicle accidents is much higher for younger drivers than for older drivers. Data for 1995 revealed that 22 per cent of injury accidents for males aged 17–19 involved no other vehicle but the driver's own.

Risky driving habits such as driving too fast in combination with a lack of the appropriate skills play a role in this type of accident. After novice drivers have obtained their licence, they must learn to match their capabilities with the task demands. If drivers underestimate the task demand or overestimate their capability, the perceived level of task difficulty will be less than is objectively the case. Accuracy of driver perception is referred to as the driver's calibration accuracy (Fuller 2011). Inexperienced drivers are prone to this type of error (De Craen 2010; Fuller et al. 2008). The authors argue that the learning process of calibration increases greatly under the influence of a high quality of feedback. It is crucial that inexperienced drivers acquire the correct behaviours and that errors generate adequate feedback. However, as traffic is often 'very forgiving', this condition is not met. Often, there are no adverse events telling a driver that they have encountered or created a situation in which demand exceeds capability. The absence of realistic feedback creates a condition that stimulates an illusion of control and self-enhancement (Rothengatter 2002, 2005). Drivers begin to consider themselves more skilful than they actually are. This bias in perception has been related to self-enhancement, optimism bias and illusion of control. There is a tendency in drivers, and especially in young drivers, to expect positive outcomes of a situation, which arises from a persistent overestimation of the degree of control drivers have over events and their outcomes.

Motivational Features

Drivers differ in their individual dispositions to adopt certain behaviours – for example the High Unsafe Speed Accident Reduction (HUSSAR) project. When asked about their preferences for speed choice, the answers showed a wide

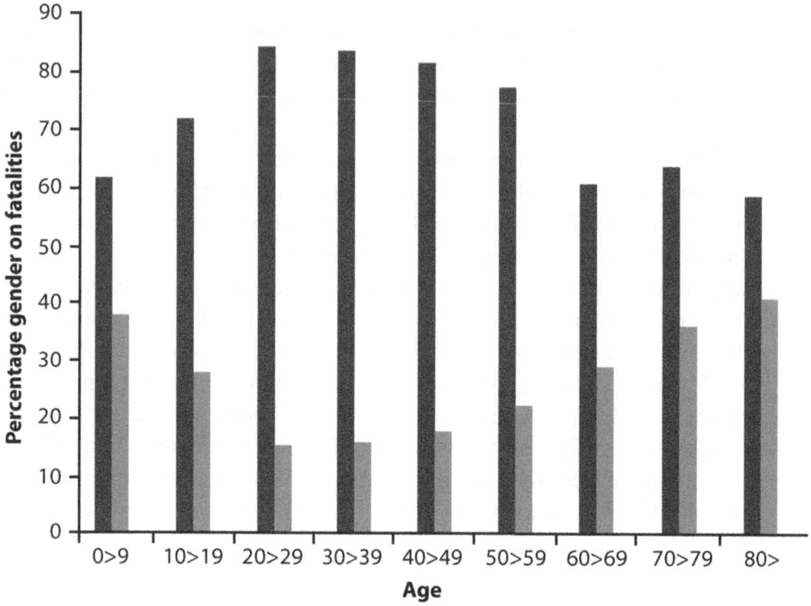

Figure 10.2 Distribution of fatalities by gender and age group (in percentages) (from EU-19, 2006; statistics compiled from ERSO)

variation in preferred speed (Stradling et al. 2008). From these and other studies in the context of the HUSSAR project, a distinction was made between four groups of drivers: low risk threshold, high risk threshold, opportunistic and reactive (Fuller et al. 2008). Young drivers are typically members of the group with a high risk threshold. Risk threshold refers here to the upper limit of task difficulty a driver will accept. There is a subgroup of young drivers who deliberately take risks and who see high-risk driving as a demonstration of superior control skills associated with social status and reinforced by peer influences (for example Fuller 2002b). It is estimated that about 15 per cent of young male drivers fall into this category. Their risk taking is part of their lifestyle and other deviant behaviour. The subgroup 'drives at higher speeds; commit more, and more extreme, speed limit violations and other forms of dangerous driving behaviour; and have more convictions' (Fuller 2011: 19). Other studies support the finding that young male drivers are more inclined to violations than adult or female drivers are (Yagil 2005; Mesken, Lajunen and Summala 2002) – see Figure 10.2.

Effects on Driver Behaviour

Novice drivers generally make more errors in traffic. For this group it is not only the complexity of the driving task itself that is the important issue, but also the extent to which novice drivers make the task difficult for themselves. There is a tendency for fast driving, keeping headway distances that are too short and driving under difficult conditions – for example, in older vehicles, during darkness or with adverse visibility.

In a research study by Lajunen and Summala (1997) a distinction was made between skill orientation and safety orientation of young male drivers. The authors collected extensive questionnaire data of 195 male drivers on driving skill and personality and observed driver behaviour of 58 male subjects. From the results it appeared that skill-oriented drivers like driving and therefore drive more and are also involved in more accidents and traffic violations. The authors hypothesize that skill-oriented drivers also have at tendency to satisfy their need for sensation in traffic by speeding (Lajunen and Summala 1997: 287). The safety orientation of drivers was negatively related to driving aggression and sensation-seeking. The results suggest that those drivers who (over)trust their skills drive faster and accept higher acceleration. The more safety-oriented drivers recognize the risks of high speed and, therefore, use lower speeds. In general it can be said that novice drivers often fail in finding a good balance between traffic task complexity and their own competence, and thereby increase their exposure to risk (Kuiken and Twisk 2001; De Craen 2010). The balancing of task demands and skills has been referred to as 'calibration.

Calibration is an ongoing process. A crucial element in the calibration model is that balancing and adaptation can take place at different hierarchical levels of the driving task. A driver can for instance choose to reduce speed (adaptation at a lower level) or not to drive at night (adaptation at a higher level). Overestimation

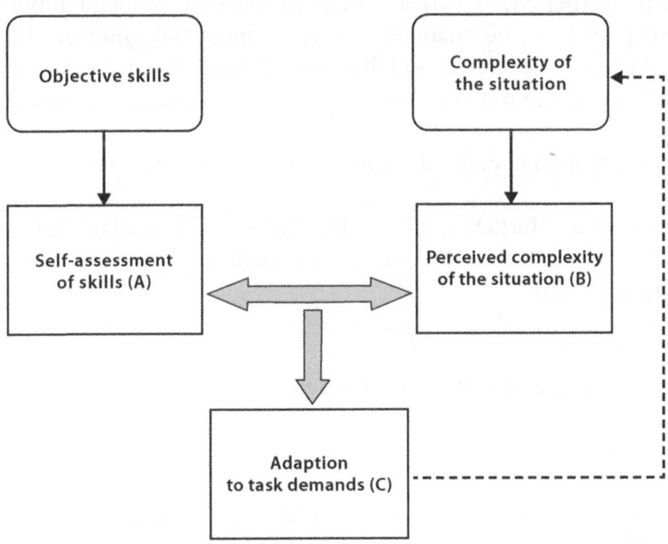

Figure 10.3 Model of the calibration process (after De Craen 2010)

of skills is a negative factor because it results in less or insufficient adaptation to task demands (De Craen 2010).

Results of a study by De Craen (2010) support for the notion of calibration and the separate elements of the task (see Figure 10.3). Young drivers overestimate their driving skills more than experienced drivers, and inadequate self-assessment is connected to insufficient adaptation to task demands. Compared to other age groups young drivers are more likely to engage in secondary behaviours (for example, making a telephone call while driving (Sayer et al. 2005). At the same time there are indications that young, inexperienced drivers, as a group, do not compensate for their (deficient) level of automated driving by decreasing task demands (Twisk 1995).

Implications for Design

Fuller (2002b) proposes a number of engineering measures. These measures address the problem of information overload, a problem that novice drivers are particularly prone to. The measures are:

- Ensure that drivers' attention is reliably attracted to the occurrence and nature of hazards ahead, for instance by using rumble strips.
- Give clear guidance as to what to do (where to go, what speed); give the information well in advance.
- Use traffic-calming measures such as humps, ramps and so on to bring the speed behaviour under control.
- Help inexperienced drivers 'help themselves' through modifying their perceptions of the roadway so as to induce heightened attention and vigilance or even perceptual illusions of speed. For instance, use transverse lines across the roadway with progressively decreasing intervals between each.
- Avoid attentional conflicts in high stimuli environments.

These solutions are further qualified by Fuller by indicating that such design measures for young drivers are likely to benefit all road users, and especially drivers who are elderly.

Emotions and Aggressive Driving Behaviour

Emotions in Traffic

Emotions play an important role in everyday life, including how we behave in traffic. Emotions are regarded as abstract, high-level descriptions of the relationship between a person and their environment (Groeger 2000). Studies reviewed by

Mesken, Hagenzieker and Rothengatter (2008) show that emotions and moods may affect driving-related performance in a number of ways. It becomes clear that especially feelings of anger and hostility seem to affect general task performance. These emotions are also related to violations and aggressive and risky driving. From a literature review Mesken et al. (2007) point out that the traffic context is an environment in which emotions occur very frequently. These results confirm the findings from Underwood et al. (1999). In this study drivers filled in driving logs and reported incidents involving anger in 21.5 per cent of all journeys.

People differ in the type and frequency of emotional experience. Happiness was the most frequent emotion shown while driving (diary study by Levelt 2003). Levelt found that of all reported emotions, 54 per cent was happiness, 22 per cent was anger and 8 per cent was fear. Outcomes of the study by Mesken et al. (2007) were that anxiety was the most frequent on-road emotion, followed by anger and happiness. Anxiety was mostly associated with safety-related events, for which a traffic situation is responsible. Anger occurred mostly when progress was blocked. The responsible agent was mostly another car user. But there is also a relation of anger with personality variables. Vassallo et al. (2007) found, for instance, that high levels of antisocial behaviour and aggression, and low levels of empathy, contributed to young drivers' involvement in risky driving and speeding violations. Machin and Sankey (2008) explored the combined effects of specific risk perception variables (worry and concern, likelihood of accident, efficacy, aversion to risk-taking) and specific personality variables (anxiety, anger, excitement-seeking, altruism, normlessness) on driving behaviour of young drivers. The study shows that, in addition to being inexperienced, young drivers with higher levels of excitement-seeking, lower levels of altruism, greater perceived likelihood of accident and a lower aversion to risk-taking are more likely to report greater speeding. Greater speeding was also associated with greater aggression. A greater level of excitement-seeking was associated with a lower aversion to risk-taking, which in turn negatively predicted speeding. The authors conclude that a driver's attitudes about the social acceptability of speeding or risky driving may be the strongest influence on how likely that driver is to speed or take risks while driving. Although relevant (for example for educational measures), these findings are of limited value for designing safe road systems.

Attribution

Of particular interest are the factors that evoke the impression of others' aggression. Some studies have attempted to attribute the incidence of aggression to road conditions and congestion, but that may well be beside the point. It is more likely that behaviour of others is interpreted as aggressive when road conditions or congestion place a high demand on or potential threat to drivers, while that same behaviour is interpreted as merely instrumental in less demanding circumstances. It is also likely that behaviour that is judged instrumental by the driver is considered aggressive by others (Rothengatter 2005).

Drivers differ in their tendency to become angry in traffic (for example Deffenbacher et al. 2003a, 2003b). We do not precisely understand why or to what degree driving leads to anger, but attribution is part of the process. Attribution is concerned with ways in which we understand and interpret the behaviour of others. From a study of experienced drivers taking a driving test, Groeger (2000) observed that the attributional style for positive or negative events influences our mood and the way we appraise ourselves and others. Matthews and Norris (2002) investigated the relation between hostile attribution and aggression. Findings show that whereas in normal driving and traffic conditions there was no relation between attribution and self-reported aggressive behaviour, in ambiguous situations drivers with a more hostile attribution were more aggressive than others. Gender and age differences in driving anger and aggression are minimal or non-existent.

However, people may be prone to anger; they may have a negative attributional style; they may have a negative attitude towards speed limits or consider themselves better than others; or they may be impatient, in a hurry – but most drivers seem to have sufficient skills to adapt to prevailing circumstances, which in this case means ignore their trait and state and act responsively. Road rage or other pathological behaviour occurs where this adaptive skill fails (Rothengatter 2005).

Aggressive Behaviour

Emotions in traffic can lead to aggressive and risky driving (Deffenbacher et al. 2003a, 2003b; Lajunen and Parker 2001):

> Some angry drivers aggress (e.g. yell at another driver or intimidate with their vehicle) and act in other ways (e.g. speed or drive recklessly), whereas other angry drivers aggress little and drive fairly normally. Nonetheless, both types of angry drivers experience strong angry emotionality and upset (e.g. mad, angry or furious) and accompanying physical arousal (Deffenbacher et al. 2003b: 334).

Various studies have shown that the driving context is a relevant factor in aggressive behaviour. Chapman et al. (2000) demonstrate that the likelihood to act on anger is influenced by the context: in driving conditions it is more likely that people react on their anger with aggression than in non-driving situations.

Several studies investigated the relationship between mood states and performance. Results suggest that anger/aggression is related to a decrease in task performance, as rated by a driving instructor, to increased acceleration after an impeding event and to a less cautious driving style (Mesken et al. 2008). A study by Arnett, Offer and Fine (1997) has shown that respondents exceeded the speed limit to a greater degree when angry than when in any other emotional state. On-the-road-studies using self-reported (observed) behaviour and physiology (for example, Mesken et al. 2007) also provided substantial support for the hypothesis that there is a relation between anger and aggressive violations. Wundersitz and Burns (2008) found evidence that the use of driving to release tension was associated with traffic offences and collision involvement, particularly among males.

Anger is also related to both highway code and aggressive violation (Lajunen, Parker and Stradling 1998) and general (verbal and physical) aggressiveness on the one hand, and to aggressive reactions on the other hand (Lajunen and Parker 2001). Parker, Lajunen and Summala (2002) demonstrate that those behaviours that provoke most anger also provoked the most extreme reactions. Other studies show a consistent pattern of high-anger drivers reporting more aggressive and more risky behaviours on the road (Deffenbacher et al. 2003a, 2003b).

Emotion and Concurrent Tasks

Concurrent tasks (telephone use, integrated systems manipulation or radio listening etc.) have an impact on drivers' attention. Most people spend most of their driving time listening to music or to the radio. Pêcher, Lemercier and Cellier (2009) examined the effects of emotions on drivers' attentional behaviour and used music to convey emotion. Results show that the driving speed is dependent on the emotional value of the music. Happy music is associated with an important decrease of mean speed, whereas it slightly increases with sad music. Driving with neutral music represented a baseline situation where drivers remained focused on their primary task, driving. The authors propose a new classification of emotions in different driving situations. Negative emotions such as anger would lead to aggressive behaviour: faster speeds, etc. Negative emotions such as sadness would lead to a more passive attitude. Positive emotions such as joy and happiness would engage drivers and distract them with risk-taking. Further research may address the question of what exact function is influenced by emotion.

Implications for Design of Road Systems

After extensive research, Mesken (2006) argues that because emotions mostly occur in interactions with other road users, measures to reduce congestion are not efficient to prevent anger on the road. A better strategy is to make the traffic system more predictable, as is proposed with the concept of 'sustainable safety' (see also Chapter 2). Sustainable safety of the road system limits aggressive behaviour primarily because it reduces the number of interactions between road users. From this perspective, unambiguous situations should be avoided and roads should be designed consistently in order to support the expectations of road users.

Summary and Conclusions

The key message to those considering designing an environment for drivers is that people are not only and simply information processors; they are not always rational and their performance may vary in different circumstances (O'Connell

2002). Sometimes drivers will simply follow what other drivers are doing; at other times they are in a hurry or involved in other tasks. People have their own motives and are optimizing their own goals. Also, it should be realized that age-related behavioural changes occur at different levels and rhythms, leading to high variability among people. Engineers have the challenging task of taking account of drivers' capabilities and individual as well as momentary differences in these capabilities.

It is expected that the participation of older people in traffic will strongly increase over the coming years. Driving difficulties and weaknesses of older drivers concern motion perception and contrast sensitivity; overlooking other road users while merging and changing lanes; overlooking traffic signs and signals; and increased reaction times. Older drivers will benefit from measures that reduce task demands and increase decision time.

Inexperienced drivers have less developed control skills and make more errors; they are particularly prone to information overload. This group, especially the male drivers, has a strong tendency to overestimate their own skills and their ability to correct a situation. The illusion of control and optimism bias is stimulated by the absence of realistic feedback. These drivers will benefit from traffic-calming measures and measures that address the problem of information overload (for example, measures that ensure that drivers' attention is reliably attracted to the occurrence and nature of hazards ahead).

Feelings of anger and hostility seem to affect general task performance. These emotions are also related to violations and aggressive and risky driving. Reducing the number of interactions between road users and avoiding ambiguous situations are recommended strategies to address this problem.

The rule of variance and individual differences dealt with in this chapter leads to a few basic notions for designing safe roads, namely:

1. Road systems should have a large tolerance for variance in behaviour.
2. Systems should be designed in such a way that people find it easy to use them. It is recommended to avoid difficult situations on the road that cause stress and anxiety.
3. Use clear and multiple warning signs well in advance of an incident or change of the road (such as roadworks or detours) to create sufficient time and opportunity to detect the change or incident. This is especially important for older drivers and fatigued drivers, who drive under unfavourable circumstances – for example, at night or when it is raining.
4. Do not use rules that cannot be enforced and that can be used in different ways by road users. This will lead to variance in behaviour which is associated with higher risks. An example where this sometimes occurs is in putting up several signs for the maximum speed on motorways (varying from 60 km/h, 70 km/h, 80 km/h to 90 km/h). Some drivers will give up trying to drive at the right speed.

References

Arnett, J., Offer, D. and Fine, M.A. 1997. Reckless driving in adolescence: 'State' and 'trait' factors. *Accident Analysis and Prevention*, 29: 57–63.

Brouwer, W., Rothengatter, T. and Van Wolffelaar, P. 1988. Compensatory potential in elderly drivers. In: J.A. Rothengatter and R.A. de Bruin (eds), *Road User Behaviour*. Assen: Van Gorcum.

Chapman, P., Evans, J., Crundall, D. and Underwood, G. 2000. Anger and aggression in driving and non-driving contexts. In: T. Rothengatter and R.D. Huguenin (eds). *Proceedings of the International Conference on Traffic and Transport 2000*. Oxford: Elsevier.

Clark, D., Ward, P. and Truman, W. 2002. In-depth accident causation study of young drivers (TRL 542). Crowthorne: Transport Research Laboratory.

Commission of the European Communities. 2006. *The Demographic Future of Europe: From Challenge to Opportunity* (EU25. COM 571). Brussels.

Davidse, R.J. 2007. *Assisting the Older Driver: Intersection Design and In-Car Devices to Improve the Safety of the Older Driver*. SWOV-Dissertatiereeks. Leidschendam: SWOV Institute for Road Safety Research.

De Craen, S. 2010. *The X-Factor: Stichting Wetenschappelijk Onderzoek Verkeersveiligheid SWOV*. Leidschendam: SWOV Institute for Road Safety Research.

Deffenbacher, J.L. Deffenbacher, D.M., Lynch, R.S. and Richards, T.L. 2003a. Anger, aggression and risky behaviour: A comparison of high and low anger drivers. *Behaviour Research and Therapy*, 40: 717–37.

Deffenbacher, J.L., Lynch, R.S., Filetti, L.B., Dahlen, E.R. and Oetting, E.R. 2003b. Anger, aggression, risky behaviour, and crash-related outcomes in three groups of drivers. *Behaviour Research and Therapy*, 41: 333–49.

De Waard, D. 2002. Mental workload. In: R. Fuller and J.A. Santos (eds), *Human Factors for Highway Engineers*. Oxford: Elsevier, 161–75.

Eurostat 2004. Available at http://epp.eurostat.ec.europa.eu/portal/page/portal/population/introduction [accessed: 19 April 2012].

Fuller, R. 2001. Learning to drive. In: P.E. Barjonet (ed.), *Traffic Psychology Today*. Dordrecht: Kluwer Academic Publishers.

Fuller, R. 2002a. Human factors and driving. In: R. Fuller and J.A. Santos (eds), *Human Factors for Highway Engineers*. Oxford: Elsevier, 77–97.

Fuller, R. 2002b. The psychology of the young driver. In R. Fuller and J.A. Santos (eds), *Human Factors for Highway Engineers*. Oxford: Elsevier, 241–54.

Fuller, R. 2011. Driver control theory: From task difficulty homeostasis to risk allostasis. In: B. Porter (ed.), *Handbook of Traffic Psychology*. Amsterdam: Elsevier, 13–26.

Fuller, R., Bates, H., Gormley, M., Hannigan, B., Stradling, S., Broughton, P.S., Kinnear, N. and O'Dolan, C. 2008. *The Conditions for Inappropriate High Speed: A Review of the Research Literature from 1995 to 2006*. Road Safety Research Report No. 92. London: Department for Transport.

Groeger, J.A. 2000. *Understanding Driving: Applying Cognitive Psychology to a Complex Everyday Task*. Hove: Psychology Press.

Groeger, J.A. 2006. Youthfulness, inexperience and sleep loss: The problems young drivers face and those they pose for us. *Injury Prevention*, 12(Suppl. 1): i19–24.

Kuiken, M.J. and Twisk, D.A.M. 2001. *Safe driving and the Training of Calibration: A Literature Review*. SWOV Report R-2001-29. Leidschendam: SWOV Institute for Road Safety Research.

Lajunen, T., Parker, D. and Stradling, S. 1998. Dimensions of driving anger, aggressive and highway code violations and their mediation by safety orientation in UK drivers. *Transportation Research F: Traffic Psychology and Behaviour*, 1: 107–21.

Lajunen, T. and Parker, D. 2001. Aggressive people, aggressive drivers? A study of the relationship between self-reported general aggressiveness, driver anger and aggressive driving. *Accident Analysis and Prevention*, 33(2): 243–55.

Lajunen, T. and Summala, H. 1997. Effects of driving experience, personality, drivers' skills and safety orientation on speed regulation and accidents. In: T. Rothengatter and E. Carbonell Vaya (eds), *Traffic and Transport Psychology: Theory and Application*. Oxford: Elsevier, 283–96.

European Road Safety Observatory (ERSO). Available at: http://www.erso.eu

Levelt, P.B.M. 2003. *Praktijkstudie naar emoties in het verkeer*. SWOV R-2003-8. Leidschendam: SWOV Institute for Road Safety Research.

Machin, M.A. and Sankey, K.S. 2008. Relationships between young drivers' personality characteristics, risk perceptions, and driving behaviour. *Accident Analysis and Prevention*, 40(2): 541–47.

Matthews, B.A. and Norris, F.H. 2002. When is believing 'seeing'? Hostile attribution bias as a function of self-reported aggression. *Journal of Applied Social Psychology*, 32: 1–32.

Mesken, J. 2006. *Determinants and Consequences of Drivers' Emotions*. SWOV-Dissertatiereeks. Leidschendam: SWOV Institute for Road Safety Research.

Mesken, J., Lajunen, T. and Summala, H. 2002. Interpersonal violations, speeding violations and their relation to accident involvement in Finland. *Ergonomics*, 45(7): 469–83.

Mesken, J., Hagenzieker, M. and Rothengatter, T. 2008. A review of studies on emotions and road user behaviour. In: L. Dorn (ed.), *Driver Behaviour and Training*, vol. III. Aldershot: Ashgate, 91–107.

Mesken, J., Hagenzieker, M.P., Rothengatter, T. and de Waard, D. 2007. Frequency, determinants, and consequences of different drivers' emotions: An on-the-road-study using self-reports (observed) behaviour, and physiology. *Transportation Research F*, 10: 458–75.

O'Connell, M. 2002. Social psychological principles: The group inside the person. In: R. Fuller and J.A. Santos (eds), *Human Factors for Highway Engineers*. Oxford: Elsevier.

Olson, P.L. and Sivak, M. 1986. Perception-response time to unexpected roadway hazards. *Human Factors*, 28(1): 91–6.

Parker, D., Lajunen, T. and Summala, H. 2002. Anger and aggression in three European countries. *Accident Analysis and Prevention*, 34(2): 229–35.

Patten, C.J.D., Kircher, A., Oslund, J., Nilsson, L. and Svenson, O. 2006. Driver experience and cognitive workload in different traffic environments. *Accident Analysis and Prevention*, 38(5): 887–94.

Pêcher, C., Lemercier, C. and Cellier, J. 2009. Emotions drive attention: Effects on driver behaviour. *Safety Science*, 47: 1254–9.

Quimby, A.R. and Watts, G.R. 1981. Human factors and driving performance. Laboratory Report 1004. Crowthorne: Transport Research Laboratory.

Rothengatter, J.A. 2002. Drivers' illusions – no more risk. *Transportation Research F*, 5: 249–58.

Rothengatter, T. 2005. Traffic psychology and road safety: Separate realities. In: G. Underwood (ed.), *Traffic and Transport Psychology: Theory and Application*. Oxford: Elsevier, 595ff.

Sayer, J.R., Devonshire, J.M., Flannagan, C.A. 2005. *The Effects of Secondary Tasks on Naturalistic Driving Performance*. Report UMTRI-2005-29. Ann Arbor: The University of Michigan, Transportation Research Institute.

Shiffrin, R.M. and Schneider, W. 1977. Controlled and automatic human information processing: II. Perceptual learning, automatic attending, and a general theory. *Psychological Review*, 84(2): 127–90.

Simões, A. and Marin-Lamellet, C. 2002. Road users who are elderly: Drivers and pedestrians. In: R. Fuller and J.A. Santos (eds), *Human Factors for Highway Engineers*. Oxford: Elsevier.

Stradling, S., Broughton, P.S., Kinnear N., O'Dolan, C., Fuller, R., Bates, H., Gormley, M. and Hannigan, B. 2008. *Understanding Inappropriate High Speed: A Quantitative Analysis*. Road Safety Research Report No. 93. London: Department for Transport.

Twisk, D.A.M. 1995. *Factors Contributing to the High Accident Liability of Novice Drivers and the Role of Driver Training*. Report D-95-9. Leidschendam: SWOV Institute for Road Safety Research.

Underwood, G., Chapman, P. Wright, S. and Crundall, D. 1999. Anger while driving. *Transportation Research F*, 2: 55–68.

Vassallo, S., Smart, D., Sanson, A., Harrison, W., Harris, A., Cockfield, S. and McIntyre, A. 2007. Risky driving among young Australian drivers: Trends, precursors and correlates. *Accident Analysis and Prevention*, 39(3): 444–58.

Wegman, F. and Aarts, L. 2006. *Advancing Sustainable Safety: National Road Safety Outlook for 2005–2020*. Leidschendam: SWOV Institute for Road Safety Research.

Wickens, C.D., Gordon Becker, S.E., Liu, Y. and Lee, J.D. 2004. *An Introduction to Human Factors Engineering*, 2nd edition. Upper Saddle River, NJ: Pearson Prentice Hall.

Wundersitz, L. and Burns, N. 2008. Personality and attitudinal predictors of traffic offences among young drivers: A prospective analysis. In: L. Dorn (ed.), *Driver Behaviour and Training*, vol. III. Aldershot: Ashgate, 51–63.

Yagil, D. 2005. Drivers and traffic laws: A review of psychological theories and empirical research. In: G. Underwood (ed.), *Traffic and Transport Psychology: Theory and Application*. Oxford: Elsevier.

Chapter 11
Communicating with the Road User

Maria Kuiken

Communication can contribute to behavioural change. The effectiveness of communication strategies depends on many factors.

Theoretical Background

Behaviour occurs in a context. The physical environment is important to our thoughts, feelings, behaviour and well-being (Steg and Gifford 2008). Consequently, it makes sense that road authorities should, as a first principle, make roads and road environments 'self-explaining'. Only when situations are not self-explaining will additional communication be required. That oversimplifies matters of course, because there are many traffic situations and occasions where additional communication to road users is needed. This may be in special circumstances (for example roadworks); in situations where we want to influence decisions of drivers at a strategic level (for safety or throughput); and when drivers violate rules and regulations. It is also necessary to communicate with road users when infrastructure, regulations or procedures have changed and *new* behaviour is required, even when the changes lead to self-explaining roads (SERs). The reason for this is that much of driver behaviour is automatic or habitual behaviour. This chapter addresses the question how to effectively communicate with the driver. When should information be given and in what way? Should we use mass media campaigns, roadside messages or personalized information?

From the behavioural science literature we know that various interventions can be employed to change behaviour. The behavioural analysis approach distinguishes *antecedent* and *consequence* strategies for change. Antecedent strategies are aimed at directing behaviour necessary to obtain desired outcomes. They precede behaviour, and change environmental stimuli that occur before target behaviour. Examples are the provision of information, or self-explaining roads. Consequence interventions are aimed at changing the consequences of behaviour, and follow behaviour and control behaviour. Examples are feedback or fines. Desirable behaviour is rewarded and/or undesirable behaviour is punished. Interventions, also in traffic, are either based on an antecedent or a consequence strategy, or on both.

Many theories on human behaviour assume that people are predominantly rational and make conscious decisions: that is, it is assumed that people think before they act. However, much of our behaviour is habitual, automatic and/or unconscious. Many aspects of the driving task, especially those performed at operational and tactical levels, are performed without thinking. When habits are strong, people no longer make reasoned choices. Clearly, it is very useful for the driver to develop habits because this saves cognitive effort. On the other hand, it is also clear that there are many situations where conscious decision making will and should take place and where road authorities require a road user to behave differently than they habitually do. When behaviour is largely automatic, changing it will require more than just providing information. Messages on the roadside or from campaigns might not even attract people's attention because they unconsciously filter the information out. The challenge is then to attract the attention of road users and make them conscious of the situation. This can be accomplished for instance by using antecedent interventions. Next, the driver has to be persuaded, or forced, to adopt the desired behaviour. Repetition and consequence interventions – for example, positive feedback – can eventually lead to the formation of new habits. In other words, it is crucial that interventions are focused and adapted to the specific requirements of drivers and their context.

Theories of social influence can assist road authorities in making behavioural interventions more effective. In his book, *Influence: Science and Practice*, Cialdini (2001) identified six social influence principles. Lehman and Geller (2008) discussed these principles and provided a concise overview of the various strategies and tools to be used for behaviour-focused interventions. The three approaches – consequence strategy, antecedent strategy and strategy using principles of social influence – are discussed below.

Behavioural Interventions: Consequence Strategy

Consequence interventions are based on the assumption that behaviour can be changed by changing the positive or negative consequences of behaviour, and some behaviour analysts consider consequences to be the main determinant of voluntary behaviour (Lehman and Geller 2008). This type of intervention is generally effective in promoting behaviour change. The three basic consequence strategies are: penalties, rewards and feedback.

Penalties

Penalties involve linking negative consequences to undesirable behaviour. This approach is often used in traffic. A major factor determining traffic law compliance is the subjective probability of detection when committing an offence. For this reason, this type of intervention requires extensive enforcement in order to be effective, especially when laws and traffic rules are involved. Efficient traffic law enforcement must maximize the objective probability of being caught. This probability can be increased by using (semi)automatic enforcement systems that detect and register both the violation and the violator, for example speed cameras. Speed measurement coupled to a camera registering the licence plate is in use in most European countries. Similar devices are being used for registering red light violations. The penalties are usually preceded by antecedent messages that announce the penalty and preferably the (increased) enforcement of the specified behaviour.

This strategy is effective and important for promoting traffic safety (for example, Rothengatter 1991). Essential to the successful implementation of such systems is that it be legally accepted that the vehicle owner is culpable for the offence committed with their car, irrespective of whether they were actually driving at the time the offence was committed.

A potential problem is that most people react to penalties with negative emotion and attitudes. They only engage in the desirable behaviour because they wish to avoid negative consequences, and only do so as long as the penalties are in place. It does not contribute to a safety culture or a positive decision to behave safely in any traffic situation. Another problem is that legal procedures may limit the efficacy of the penalties because of the long delay in informing a driver that they have been registered as having committed a traffic law violation (Rothengatter 1991).

Rewards

Rewards are positive consequences following desired behaviour. The positive consequences could be money or privileges. The basic notion is that the rewards may serve as an extrinsic motivator to perform the desired behaviour. The positive consequences are often preceded by antecedent messages announcing the availability of the reward upon completion of a specified behaviour. This has a positive effect on the outcome (Lehman and Geller 2008). Such antecedent messages are termed 'incentive'.

There are some examples of using rewards in the context of travel and traffic. For example, driving without collisions leads to lower insurance premiums. In the Netherlands, large-scale pilot studies are conducted to investigate the effects of financial rewards in road use. In one of these studies road users receive money each time they decide not to use a certain road at a certain time of day. The strength of such systems is that even small rewards can have substantial effects. The weakness is that rewards provide extrinsic feedback, which can undermine the intrinsic motivation of a person to behave responsibly or safely. The moment the rewards stop, people tend to fall back to their original habitual behaviour.

Feedback

Feedback means providing information about the consequences of behaviour. The provision of feedback can be an effective strategy to reach a certain goal. It is particularly important for learning new behaviour. Feedback is most effective when it is provided immediately after the behaviour occurred, and when people can clearly link the consequences to their behaviour. This is also important when using penalties, for example, fines after having committed an offence. Providing the feedback at the time the offence is committed will greatly increase the effect of the penalty. Feedback provisions have been a powerful means to increase compliance (for example, Riedel, De Bruin and Rothengatter 1985).

Two types of feedback are very common in traffic: normative and comparative feedback. An example of normative feedback is information on roadside electronic boards when a driver either stays within the speed limit or exceeds it

Comparative feedback is feedback about a person's performance relative to the performance of others. An example is the roadside posters which inform the public about the percentage of road users in a particular timeframe (a day or a week) who did not drive faster than the speed limit. Comparative feedback may be counter-effective when the majority of others do violate traffic rules. The behaviour of others indicates which behaviour is most appropriate or effective in a given situation, so if most people violate traffic rules people may think this is the appropriate behaviour.

Many studies demonstrated that consequence strategies, such as discussed above, can in principle be effective. It is also clear that the positive effect of rewards, penalties or feedback may disappear as soon as the intervention is discontinued (Abrahamse et al. 2005). This is because the feedback (penalty or reward) is extrinsic or artificial. Consequence interventions usually do not take account of the fact that there are immediate social influences on the driver: for example, economic factors, enforcement of legislation and aspects of our cultural and physical surroundings. These influences may overrule the cognitive processes and can cause effects to disappear when the intervention stops.

After reviewing literature on modification programmes and information measures, Lund and Aarø (2004) argue that a combination of information and preventive measures should be used, dependent on characteristics of the driver and the context. They indicate that measures are probably more effective when they not only reach individuals but also succeed in influencing social norms and public opinion. The conclusion of these authors is that:

> It is a challenge to identify cultural dimensions that may have an impact on safety practices. Such factors have to be taken into account when planning preventive action. In the long term, changes in such cultural factors may prove to be a prerequisite for effectively promoting safety within a country or culture (Lund and Aarø 2004: 314).

And:

> Preventive action should not be initiated and designed 'behind the backs' or 'over the heads' of the general population. We are not in the business of social manipulation or the art of engineering. Attitude change approaches are an integral part of a process that is concerned by all sectors and levels of society. This is consistent with a modern view of primary prevention and health promotion (Lund and Aarø 2004: 316–17).

Once feedback is given it is important that it is given frequently and systematically so that drivers can acquire the normative, safe procedures of driving. The absence of feedback may, on the other hand, weaken the fragile association between actions and their consequences, which may result in inappropriate behaviour and 'slipping of norms'.

Behavioural Interventions: Antecedent Strategy

Antecedent interventions influence one or more elements in the environment prior to the performance of behaviour. Many antecedent strategies are presumed to work because they announce the availability of consequences associated with the desired behaviour. The most common antecedent strategy is the provision of information. Another approach is the use of modelling.

Media Campaigns

Providing information is one of the strategies to promote certain behaviour. The information can be conveyed in several ways: by education, training, mass media campaigns, by the use of personalized information on road signs or by using prompts. In education, a target group is given information and is then expected to perform the required behaviour. Training goes a step further and assists people in acquiring the necessary skills and attitudes to perform the required behaviour. Mass media campaigns are mostly used for large-scale interventions. The communication campaigns consist of purposeful attempts to inform, persuade or motivate people with a view to changing their beliefs and/or behaviour in order to improve road safety (CAST project;[1] Delhomme et al. 2009). There are many examples of mass media campaigns used to increase traffic safety. Mostly a communication campaign is combined with other interventions such as enforcement, legislation and education. Sometimes they accompany engineering improvements or a change in road design. There are also examples of integrated approaches, such as the umbrella campaign brand in the United Kingdom called 'THINK'. The objectives were improved road user behaviour and acceptance of engineering and enforcement initiatives that increase road safety. The outcomes were positive (Delhomme et al. 2009).

The GADGET project – Guarding Automobile Drivers through Education and Technology – performed a meta-analysis that provides insight into the key elements of successful road safety campaigns (Delhomme et al. 1999). The main conclusion was that some road safety campaigns do significantly reduce accidents. The effects were attributed to all components of the campaigns, including activities such as enforcement, rewards, legislation and educational programmes, in addition

1 CAST – Campaigns and Awareness-raising Strategies in Traffic Safety – a project co-financed by the European Commission.

to the media campaign itself (Delhomme et al. 2009). The effects were greater for campaigns combined with enforcement, legislation or rewards. Campaigns carried out alone did not seem to produce any significant reduction in the number of collisions during the campaign period or after the campaign.

The GADGET project provided no insight into the mechanisms and variables that contribute to accident-reducing effects. The INFOEFFEKT project (Vaa et al. 2004) updated, reorganized and reanalysed the GADGET database. The main finding from this project was that campaigns against drinking and driving reduced accidents significantly. No significant effect was found for campaigns against speeding. There was also support for the findings in earlier studies that mass media campaigns alone (for example, only TV, radio and/or newspapers), without any accompanying measures, had no impact on accidents. The main recommendations from these and other studies were that the effectiveness of a road safety communication campaign is enhanced by the following key elements:

- combining communication with enforcement, education and/or legislation;
- the use of a theoretical model;
- basing campaigns on prior research;
- choosing a single theme rather than multiple themes;
- designing for a specific target group rather than addressing the whole population.

Personalized Information

Sometimes the information in communication campaigns is provided locally and specifically to users. In the literature this is called 'tailored' or 'personalized' information. This type of information, which is specific and adjusted to the individual, is potentially very effective in encouraging behavioural changes (Abrahamse et al. 2007). Tailoring is an approach frequently used in health psychology as part of interventions aimed at changing unhealthy practices into healthy ones (for example, smoking cessation). Delhomme et al. (2009) also report on an INRETS[2] study that reanalysed the GADGET data. One of the main findings was that personal influence was the only kind of information that significantly contributed to explaining why the campaigns reduced the number of collisions. For example, if letters are addressed personally to members of a given target group this can be seen as a source of personal influence.

Prompts

Prompts are messages delivered by the authorities at the place where the target behaviour occurs. The messages serve as reminders to perform the target behaviours (Lehman and Geller 2008). This is why they are only effective when people have

2 INRETS – the French National Institute for Transport and Safety Research.

already decided to perform the required behaviour. Prompts are regularly used in traffic for simple behaviours – for example, to direct road users to a certain place or to ask them to behave in a certain way. Prompts are popular because they are easy to implement, do not cost much and can be effective if implemented in the right way. Prompts are not effective if people do not accept the prompt and do the opposite of what it demands. According to Lehman and Geller (2008), conditions under which prompting strategies are most effective are when:

- The target behaviour is clearly defined by the prompt.
- The target behaviour is relatively easy and relatively convenient to perform.
- The message is displayed in close proximity of the target behaviour.
- The message is stated politely.

Modelling

Modelling provides another strategy for antecedent interventions. It is based on Bandura's learning theory (1977). Essentially, it demonstrates the recommended behaviours. Modelling is especially valuable when complex behaviours need to be learned. In traffic this strategy is only used indirectly. For instance, in media campaigns the recommended behaviour is demonstrated on television. An example is the Dutch campaign '*Rij met je Hart*' ('Drive with your heart'), where on television spots people showed tolerant behaviour to each other and were relaxed, even when another road user had misbehaved. One of the notions in this campaign was to approach traffic behaviour in a positive way and to correct the fundamental attribution error that all deviant behaviour is purposeful and consciously done. 'Aggressive behaviour' might be just the result of an error.

Another good example of modelling is the Dutch national ecodriving programme 'Het Nieuwe Rijden' ('New driving'). The programme targeted CO_2 emission reductions in traffic and transport and was a long-term strategy for the period 1999–2010. The programme was supported by a corporate communication and marketing strategy which stressed individual benefits and hedonistic elements such as safety, comfort, fun of driving and cost reductions. Energy savings and CO_2 emission reductions come as a side effect of individual benefits. Yearly evaluations of the ecodrive programme showed that a large part of the respondents were familiar with the programme. Drivers also increasingly implement aspects of the programme: 70 per cent of respondents indicate that they frequently use one or more of the recommendations that are part of the programme – for example, driving style, checking tyre pressure and in-car application for fuel savings. The effects of the programme on CO_2 emissions have been calculated and show a yearly reduction of 0.3–0.6 Mton (SenterNovem 2007).

With modelling it is assumed that the behaviour that is demonstrated will be followed when it is understandable, relevant, meaningful and rewarding to people (Abrahamse et al. 2005).

Interventions Using Principles of Social Influence

Theories of social influence can be used to strengthen the effect of antecedent and consequence strategies. These theories focus primarily on the inner conditions of people – their thoughts, feelings and attitudes. Cialdini (2001) identified six social influence principles that he called 'compliance techniques' because they increase the chance that people comply with a request. Some of the principles are particularly relevant for communication strategies in traffic, such as the principles of consistency, of social proof and of liking.

Consistency

The principle of consistency can enhance the efficiency of consequence strategies in particular. People have a strong desire to show others that they are consistent (Cialdini 2001). When setting out to change road user behaviour, using this principle means that the change process starts with small and easy acts – 'a foot in the door'. Because of the wish to be consistent, once this works one can come up with a larger request. An example: Imagine that authorities have set speed limits of 80 km/h on the motorways because of environmental reasons (CO_2 emission or noise) and they request drivers to respect these limits. The first step is that drivers become aware that the speed regulation is made for environmental reasons. Secondly, when following the principle of consistency, authorities would try to find a small act for drivers to begin with, such as displaying a sticker on the car indicating that a person is committed to environmental or safety goals. When there is some success they would continue from there.

Social psychologists found that commitment strategies are very effective in changing behaviour, especially when the commitment is public and perceived as voluntary (Cialdini 2001). It is in principle a very powerful tool because once people are strongly committed, they begin to adjust their identities to become consistent with that behaviour, and thus they are more likely to keep doing it.

Another concept attached to this is *behavioural self-perception*. This is particularly important when considering using rewards to promote certain behaviour. Lehman and Geller (2008) point out that it may be more effective if rewards are only used as tokens of appreciation. This principle is called behavioural self-perception. What counts is how the person perceives themself. The rewards should be large enough to encourage certain behaviour and thus stimulate a positive self-perception, but it should not be too large because then the reward will be an extrinsic motivation.

Social Proof

In many situations, people tend to follow the crowd. Descriptive norms are codes of conduct that inform us how the social group we belong to generally acts.

Figure 11.1 Poster in Dutch media campaign (drive cleaner, drive at 80 km/h in fifth gear)

Behavioural interventions that use the power of the group (norms) are potentially very effective.

When the desired behaviour is very uncommon, it is not recommended to provide description norm information because this may be counter-effective. For example, the Dutch campaign requesting drivers to reduce speed on certain motorways around large cities has not taken the principle of social proof into account (see Figure 11.1). The poster shows a girl with a handkerchief over her mouth because of the polluting effects of fast driving and a speed limit of 80 km/h. This suggests that there are many people who break the rule. It may be more effective to give a message of things we do well.

Violating norms and regulations leads to more violations, and also to violations in other domains. This is called the 'cross-norm inhibition effect' (see Keizer, Lindenberg and Steg 2008; Torgler 2010). This is why enforcement is important. For these behaviours, it might be better to provide personalized normative feedback.

Liking

We are more likely to do things for people when we know and like them. This principle may be of use when road authorities wish to change driver behaviour for safety reasons, for example to protect roadworkers. Making safety more personal is an example of applying the principle of liking. Using famous people to promote certain behaviour is an example of this method. Communicating with smileys and short thank you messages on the road can also have the effect of being 'more personal'.

Reciprocity Prompts

When someone does something for another person, the person will feel indebted. This is because of the principle of reciprocity. Messages announcing that road authorities have already acted on behalf of the road user may invoke reciprocity. It invites the road user to support what has already been initiated. The principle is essentially a pre-behaviour reward to encourage participation (Lehman and Geller 2008).

Designing Communication to Drivers

Delhomme et al. (2009) describe five logical steps in designing communication to drivers, namely:

1. problem diagnosis and goal setting;
2. specification of target behaviour;
3. specification of target population;
4. design of the type of intervention;
5. definition of the type of message.

From the literature it is clear that it is important that communication to drivers is specific. In general, intervention techniques are more effective inthe extent to which they specifically address those factors that contribute substantially to the particular problem behaviour (Abrahamse et al. 2005; De Vrieze, 2001; Delhomme et al. 2009). A critical factor is identifying the target group. For instance, a road safety campaign generally must be sensitive to the age group it is targeting (Fuller 2002). The next step concerns the selection of the strategy for communicating with road users (for example, antecedent and consequence strategies and the use of principles of social influence). The final step is the definition of the type of message. Information can be provided by means of roadside signs, prompts or via the media, the Internet etc. Finally, it is recommended that the intervention is evaluated.

The Process of Change

Designing communication to drivers not only requires information about effective strategies, it also demands insight into the process of change itself. This is important because it is possible to follow the five steps described above, take into account all the necessary tools for the intervention discussed earlier in this chapter and still obtain no satisfactory result with the target group. This has to do with the fact that the people whoreceive the information or intervention may be in different stages of change. The Transtheoretical Model of Change (Prochaska and DiClemente 1983) provides a useful framework by distinguishing six stages of change in people: pre-contemplation, contemplation, preparation, action, maintenance and termination.

Pre-Contemplation

The individual has no intention of changing their behaviour, and may even resist change. An example is the driver who always speeds and has not considered whether this behaviour is acceptable. When people are in this first stage of change, they basically should be made aware of the problem and their role in it. If they are not conscious of the problem, nothing will change.

Contemplation

In this stage the individual starts to become aware of the problem, with the costs and benefits of the old behaviour being equal. They are now open to new information. The message needs to be relevant and persuasive to the road user. They need to be convinced that a change (in norm or behaviour) is necessary.

Preparation

The individual starts preparations for change. Only when the message has been taken in will people prepare for action. In this stage the context becomes very important. What are the specific circumstances? Can engineering measures assist the driver in performing the required behaviour? For example, when the desired behaviour is that car drivers need to learn that they should merge later, it is helpful that the road markings have changed in a way that encourages drivers to drive longer in their own lane. In this phase it is also important that drivers know exactly what they are required to do.

Action

Change has occurred but the risk is still high that the individual will return to previous behaviour patterns. The first experience of people with the new behaviour determines whether they will continue with the new behaviour or whether they will fall back on their old behaviour. At this stage it is recommended that the road user obtain positive feedback on their behaviour (for example, a smiley).

Maintenance

The new behaviour has become habitual. To support this process it is relevant that others behave in the same way, and that if necessary there is enforcement or from time to time a campaign to 'refresh' this behaviour. Also prompts are very useful in this stage.

Termination

The new behaviour is established and the individual is not likely to return to the old behaviour.

Essential in considering change as described in this model is that the determinants of behaviour and the effectiveness of interventions differ substantially per stage. Communication with drivers requires tailored interventions and tailored communication moments. It is not just the provision of information.

Road authorities must create a knowledge framework that allows people to understand what is expected of them and why these measures are taken in this manner and not in another manner. When road authorities decide to communicate with drivers, they must realize that this takes time and a series of measures to help people understand why and how they need to perform new behaviour. People need to recognize the problem, think about it and understand the solution that is offered to them.

An example: Dutch road authorities wanted to change the merging behaviour of drivers. The idea was that people would learn to merge fluently with a method called 'RITSEN'. Firstly, people were made aware of the problem (*pre-contemplative* stage) and of the solution to that problem. There was campaign to show the correct and desirable way of merging. The goal in that stage was that people understood the concept of RITSEN and began to implement it on the road (*contemplative* stage). Next, road authorities put billboards on the road side saying '*Ritsen vanaf hier*' ('merging from here') – *preparation* stage. The authorities even went as far as going to the roads themselves and helping people to merge in the correct fashion by giving signs to individual road users etc. Also, supportive engineering measures were taken – such as a change of lane marking and shorter merging lane, thus urging the motorists to drive a bit further on the lane before merging. For maintenance reasons prompts will be necessary and it certainly will be necessary to refresh the desired behaviour from time to time.

Summary and Conclusions

Behavioural interventions have been employed with varying degrees of success. The main issues that influence the effectiveness of interventions are:

- Interventions need to be specific. This requires in-depth problem analysis and goal setting. In general, intervention techniques are more effective to the extent to which they specifically address those factors that contribute substantially to the particular problem behaviour and those factors that can be improved (Steg and Rothengatter 2008). This means that is essential to know which factors cause or inhibit behaviour before designing intervention programmes to change behaviour.

- In designing a strategy for communicating with driver a distinction can be made between an antecedent strategy and a consequence strategy. The effectiveness of both strategies can be enhanced by using social influence principles. From theories on social influence we have obtained tools and notions to improve the effectiveness of communication with drivers. The different strategies and principles can be used separately or together.
- Road authorities communicate first of all with road users through the basic design of the road (preferably self-explaining). Next, they use billboards and message signs with information or prompts. This type of communication with drivers is particularly useful for increased understanding and acceptance of (engineering) measures such as roadworks or temporary changes in the design or layout of the road. The main function of the information provided in this manner is conditional: it helps to create the right conditions for the desired behaviour to take place. In order to be effective, road authorities must attempt to create a knowledge framework that allows people to understand what is expected of them and why these measures are taken in this manner and not in another manner. People need to recognize the 'problem', think about it, understand the solution that is offered to them, and be convinced that the desired behaviour has many benefits. This takes considerably more than just the provision of information (element).
- Finally, critical information needs to offered to drivers repeatedly, 'out there in the world', rather than relying on the road user to keep it in their head (Fuller 2002: 85).

References

Abrahamse, W., Steg, L., Vlek, C. and Rothengatter, T. 2005. A review of interventions studies aimed at household energy conservation. *Journal of Environmental Psychology*, 25: 273–91.

Abrahamse, W., Steg, L., Vlek, C. and Rothengatter, T. 2007. The effect of tailored information, goal setting, and tailored feedback on household energy, energy-related behaviors, and behavioral antecedents. *Journal of Environmental Psychology*, 27: 265–76.

Bandura, A. 1977. *Social Learning Theory*. Englewood Cliffs, NJ: Prentice Hall.

Cialdini, R.B. 2001. *Influence: Science and Practice*. Boston, MA: Pearson.

Delhomme, P., Vaa, T., Meyer, T., Harland, G., Goldenbeld, C., Jarmark, S., Christie, N. and Rehnova, V. 1999. *Evaluated Road Safety Media Campaigns: An Overview of 265 Evaluated Campaigns and Some Meta-Analysis on Accidents*. EC Deliverable 4. GADGET Project. Arcueil: INRETS.

Delhomme, P., De Dobbeleer, W., Forward, S. and Simoes, A. 2009. *Manual for Designing, Implementing, and Evaluating Road Safety Communication*

Campaigns. Brussels: Campaigns and Awareness-Raising Strategies in Traffic Safety (CAST).

De Vrieze, N. 2001. Effectiveness of road safety campaigns. In P.E. Barjonet (ed.), *Traffic Psychology Today*. Dordrecht: Kluwer Academic Publishers.

Fuller, R. 2002. Human factors and driving. In: R. Fuller and J.A. Santos (eds), *Human Factors for Highway Engineers*. Oxford: Elsevier, 77ff.

Keizer, K., Lindenberg, S. and Steg, L. 2008.The spread of disorder. *Science*, 322: 1681–5.

Lehman, P.K. and Geller, E.S. 2008. Applications of social psychology to increase the impact of behaviour-focussed intervention. In: L. Steg, A.P. Buunk and T. Rothengatter (eds), *Applied Social Psychology: Understanding and Managing Social Problems*. Cambridge: Cambridge University Press, 57–86.

Lund, J. and Aarø, L.E. 2004. Accident prevention: Presentation of a model placing emphasis on human, structural and cultural factors. *Safety Science*, 42: 271–324.

Pol, B., Swankhuisen, C. and van Vendeloo, P. 2007. *Nieuwe aanpak in overheidscommunicatie: Mythen, misverstanden en mogelijkheden*. Bussum: Coutinho.

Prochaska, J.O. and DiClemente, C.C. 1983. Stages and processes of self-change of smoking: Toward an integrative model of change. *Journal of Consulting and Clinical Psychology*, 51: 390–95.

Riedel, W., De Bruin, R.A. and Rothengatter, J.A. 1985. *De invloed van intensiteit van verkeerstoezicht en variabele bebording op het snelheidsgedrag op 80 km/u wegen* [*The effects of police surveillance intensity and variable message signs on speed choice on 80 km/h roads*]. Report VK 85-19. Haren: Traffic Research Centre, University of Groningen.

Rothengatter, T. 1991. Automatic policing and information systems for increasing traffic law compliance. *Journal of Applied Behavior Analysis*, 24: 85–7.

SenterNovem, Verkeer en Waterstaat 2007. *Evaluatie het Nieuwe Rijden 2007* [*Evaluation 'Eco drive' 2007*]. NOV170/Bkr/1954.

Steg, L. and Gifford, R. 2008. Social psychology and environmental problems. In: L. Steg, A.P. Buunk and T. Rothengatter (eds), *Applied Social Psychology: Understanding and Managing Social Problems*. Cambridge: Cambridge University Press, 184–205.

Steg, L. and Rothengatter, T. 2008. Introduction to applied social psychology. In: L. Steg, A.P. Buunk and T. Rothengatter (eds), *Applied Social Psychology: Understanding and Managing Social Problems*. Cambridge: Cambridge University Press, 1–27.

Torgler, B. 2010. Comment on 'The spreading of disorder'. *Economics Bulletin*, 30: 519–23.

Vaa, T., Assum, T., Ulleberg, P. and Veisten, K. 2004. *Effects of Information on Behaviour and Road Accidents: Conditions, Evaluation and Cost-Effectiveness*. TOI Report 727/2004. Oslo: Institute of Transport Economics.

Chapter 12
Summary and Conclusions

In the previous chapters we have discussed ten general principles of how to design an adequate road environment. The underlying notion is that a good road design will more or less automatically elicit safe traffic behaviour. Our viewpoint is that road users do not necessarily intentionally want to engage in unsafe behaviour but that the road environment is designed in such a way that it allows unsafe behaviour. We do not adhere to the principle of blaming the driver for their mistakes. If the road design is suboptimal then mistakes are easily made. It is not the error of the driver; it is an error in the road system that has led to the error of the driver. If mistakes are made (and obviously humans are not infallible) then it is important that the road is forgiving – for example, an error should not automatically lead to a crash.

Our approach is also one that can be labelled as 'design for all'. We adhere to the principle that everyone with a valid driving licence should be able to drive safely on our roads. The design of the roads should not be targeted at the highly capable, young and fit adults, but at those with fewer capabilities who are older and slower. It is our responsibility to design roads in such a way that everyone is able to safely negotiate the road traffic environment. This concern is particular pressing because we have an ageing population. For example, in 2009 in the United States one in eight Americans was older than 65. It is estimated that by the year 2030 about one in five will be older than 65. It is our responsibility to take this into account when considering the road design of the future.

When considering our ten principles of road safety we can address them from a different perspective. There are principles that are clearly scientifically established and other principles that give a good direction but may be less established. The principle 'Individual information carriers should meet basic ergonomic principles to be visible, clear and understandable for the road user' is based on many years of research on minimum legibility distance, contrast sensitivity, snippets of information, conspicuity and other ergonomic 'hard fact' principles. This research is mainly based on the classic static traffic signs and their various parameters. Even though these principles are well established it is somewhat surprising that they are not necessarily applied to new electronic signs. For example, so-called graphical variable message signs (VMS) that were recently introduced in the Netherlands are not designed according to classic 'hard fact' ergonomic principles. These signs try to communicate too much information to the driver, presented in a format that is not easily understandable, giving rise to masking and crowding. In a recent study it was shown that drivers had to pass these new graphical variable message signs at least four times before they had any idea of what the signs meant. So for future reference it is important that classic basic ergonomic design standards are also taken into account when designing new electronic variable message signs.

The principle 'Individual information elements should be mutually consistent and uniform within their context' is also well established. A problem when applying this principle is that it is not always clear what consistency really is. It may appear to be consistent from the engineering point of view but from the road user's perspective it may be very inconsistent and not uniformly designed. For example, in the Netherlands we have overhead VMS indicating the maximum permitted driving speed. In some cases there is a number (say, 80 km) without a red circle around it and sometimes there is a red circle around the number (making it a prohibitory traffic sign). By law, these signs mean exactly the same (that is, you cannot drive faster than this number) but the driver may think that without the red circle it is an advisory speed, while with the red circle it is prohibitory. Another example is related to the interaction between different systems. Often the information that an in-car navigation system provides is not consistent with the information on the road signs. The navigation system may say 'take exit in 500 metres', while the sign on the road indicates an exit at 800 metres. This can be confusing to the driver and may lead to slowing down, swerving and other signs of uncertainty.

The principle 'One should take inadvertent aberrations of the system or of the road user into account. Inadvertent aberrations should not result in accidents (principle of graceful degradation)' is also well established. The principle is simple. If someone makes a mistake it is important that there is way to fix the error without problems. For example, if a driver accidentally enters a motorway via an exit (not noticing that it is an exit and not an entrance) they would ultimately end up on the motorway going against the traffic. In the Netherlands several large white arrows are painted on the exit surface such that a driver accidently entering the wrong way will notice they are driving against the arrows on the road surface. This additional warning system can ensure that a small mistake (of entering via an exit instead of an entrance) will not result in a fatal crash.

The principle 'Self-Explaining Roads (SERs) that advocate a traffic environment that elicits safe behaviour simply by its design' is very well established, rooted in a strong theoretical framework. Even though worldwide this principle is embraced by many road authorities, it is not always clear which exact measures will and will not work. The most dramatic results were reported in New Zealand, where a substantial speed reduction was found after redesigning the roads. As a general principle it should be realized that changing just a minor aspect of the road environment (putting a few signs up) will not result in significant behavioural changes. To see a real effect one needs to make radical changes, including a redesign of the road and the associated landscaping so that users will see the road as being part of different category from before. We believe that this design principle has huge potential to become the leading principle of future road design.

The principle 'Driving involves three hierarchical task levels: viz. navigation, guidance and control. For each level, task load varies depending on time and place. In road system design it is important to distinguish between all three levels' is somewhat of a more theoretical concept. Even though the concept of various task levels is theoretically well established, in practice it may not always be clear whether a driver performs a task at one or at another level. This may vary over time within and between drivers. The crucial point of this principle is that drivers may be aware of a particular measure but because the task is performed at another level the effects are small. For example, during roadworks often new yellow line markings are painted on the road surface telling drivers to follow these markings instead of the usual markings. Even though drivers may 'know' this, after a while they often start following the original road markings again because manoeuvring between road markings is a low-level automatic task which is not easily penetrable in a cognitive way.

The principle 'When approaching task overload, road users will adapt their behaviour such that the load becomes lighter. Additional tasks will be ignored or skipped. When actual task overloads occurs, rational behaviour may cease to exist and drivers may fall back to basic, rather reflexive behaviour tendencies', also has a strong theoretical background. It assumes that the driver has a limited amount of processing resources available. With increasing demands for attention from rather complex and busy road environments – with new electronic variable message signs that may or may not communicate with in-car electronics (navigation systems, crash avoidance systems etc.) – it is crucial that workload is taken into account. Even though it is clear that drivers engage in compensatory strategies when overload becomes imminent, one should realize that with every distraction safety margins erode. Also, given our principle of 'design for all' discussed earlier, one should take into account that elderly drivers in particular may quickly experience overload when confronted with complex road environments and in-car gadgets.

The principle 'When introducing particular measures it is likely that there will be adverse side effects. The driver is an integral part of the traffic system and will adapt their behaviour when necessary and possible. Any change within this system may lead to behaviour adaptation' also has a strong theoretical basis. As a general principle one should realize that when introducing a measure it is likely that some (adverse) side effect may occur. Even though some theories would claim that any safety measure would be fully counteracted, it has become clear that full risk compensation usually does not occur. Dependent on the measure, more or less compensation can be expected. Given our knowledge regarding the principle of risk compensation, it is possible to introduce measures in such a way that less compensation is expected. For example, it would be wise to not inform the public that high-friction surfacing has been introduced on the motorway. By not telling the public, the safety effects of this high-friction surface will be greater because drivers will not compensate.

This principle, 'The road user maximizes their behaviour to reach their goals, and adapts their behaviour towards that end. Non-driving activities such as telephone conversations, entertainment, relaxation and work are also included in this process', is well established but cannot be applied easily because at any moment in time and place we have no direct access to the goals of each individual driver. It is important to realize that even though in general the driver wants to go from point A to B without getting into an accident, there are other goals that may compete with this general principle. Drivers may want to engage in a telephone conversation, have lunch or send a text message. These secondary goals do play a role and can explain why drivers may get into collisions on a well-designed road, on a clear day when there is not much traffic. The 100-car naturalistic driving study (US DOT 2005) has shown that driver distractions such as eating, talking on the phone and adjusting the radio are important determinants of accidents. It is clear that drivers are not always and not all the time engaged in those tasks that are directly relevant for driving.

The principle 'Road users differ and human performance varies over time. Capabilities, social factors and momentary factors influence traffic behaviour. It may be useful to distinguish between groups. In the design of road and traffic management measures it is recommended to take characteristics of the target group into account' is related to the earlier discussed principle of 'design for all'. In the ideal situation one wants to design the road for the 'worst case' road user – those with the least skills and capabilities. Even though this principle is clear, one should realize that capabilities within an individual can also vary over time (aggression, fatigue, alcohol). For example, when drivers engage in road rage (aggressive driving) they are willing to take much more risk and engage in very dangerous manoeuvres. An Australian study estimated that about half of all traffic accidents there may be due to road rage. The effects of drunk driving are also well established. For example, in 2008 37 per cent of all fatal accidents in the US were alcohol related.

The final principle, 'Communication can contribute to behavioural change. The effectiveness of communication strategies depends on many factors', is about communicating with drivers about how they have to behave. Overall, the effects of communication campaigns on actual changes in driving behaviour are very limited. Only when such a campaign is accompanied by intensive police enforcement might one see effects that ultimately will disappear when the police no longer enforce a particular measure. Even though campaigns may change people's attitude regarding particular issues (such as speeding, drunk driving, road rage etc.) this does not imply that people actually will change their behaviour. The environment (including visible police enforcement) is a much stronger determinant of human behaviour than the cognitive intention and attitude to behave better.

Index

Entries shown in *italics* refer to figures and tables.

For Product Safety Concerns and Information please contact our EU representative GPSR@taylorandfrancis.com Taylor & Francis Verlag GmbH, Kaufingerstraße 24, 80331 München, Germany